Cambridge Elements

Elements in Philosophy of Science
edited by
Jacob Stegenga
University of Cambridge

THE SOCIAL DIMENSIONS OF SCIENTIFIC KNOWLEDGE

Consensus, Controversy, and Coproduction

Boaz Miller
Zefat Academic College and ACEPS, University of Johannesburg

Shaftesbury Road, Cambridge CB2 8EA, United Kingdom

One Liberty Plaza, 20th Floor, New York, NY 10006, USA

477 Williamstown Road, Port Melbourne, VIC 3207, Australia

314–321, 3rd Floor, Plot 3, Splendor Forum, Jasola District Centre, New Delhi – 110025, India

103 Penang Road, #05–06/07, Visioncrest Commercial, Singapore 238467

Cambridge University Press is part of Cambridge University Press & Assessment, a department of the University of Cambridge.

We share the University's mission to contribute to society through the pursuit of education, learning and research at the highest international levels of excellence.

www.cambridge.org
Information on this title: www.cambridge.org/9781009507233

DOI: 10.1017/9781108588782

© Boaz Miller 2024

This publication is in copyright. Subject to statutory exception and to the provisions of relevant collective licensing agreements, no reproduction of any part may take place without the written permission of Cambridge University Press & Assessment.

When citing this work, please include a reference to the DOI 10.1017/9781108588782

First published 2024

A catalogue record for this publication is available from the British Library.

ISBN 978-1-009-50723-3 Hardback
ISBN 978-1-108-70671-1 Paperback
ISSN 2517-7273 (online)
ISSN 2517-7265 (print)

Cambridge University Press & Assessment has no responsibility for the persistence or accuracy of URLs for external or third-party internet websites referred to in this publication and does not guarantee that any content on such websites is, or will remain, accurate or appropriate.

The Social Dimensions of Scientific Knowledge

Consensus, Controversy, and Coproduction

Elements in Philosophy of Science

DOI: 10.1017/9781108588782
First published online: December 2024

Boaz Miller
Zefat Academic College and ACEPS, University of Johannesburg
Author for correspondence: Boaz Miller, boaz.m@zefat.ac.il

Abstract: This Element is about *the social dimensions of scientific knowledge*. The first section asks in what ways scientific knowledge is social. The second section develops a conception of scientific knowledge that accommodates the insights of the first section, and is consonant with mainstream thinking about knowledge in analytic epistemology. The third section asks under what conditions we can tell, in the real world, that a consensus in a scientific community amounts to shared scientific knowledge, as characterized in the second section, and how to deal with scientific dissent. The fourth section reviews the ways epistemic and social elements mutually interact to coproduce scientific knowledge. This Element engages with literature from philosophy of science and social epistemology, especially social epistemology of science, as well as Science, Technology, and Society (STS), and analytic epistemology. The Element focuses on themes and debates that date from the start of the second millennium.

This element also has a video abstract: www.cambridge.org/Miller

Keywords: science and values, trust in science, expert testimony, scientific consensus and dissent, collective knowledge

© Boaz Miller 2024

ISBNs: 9781009507233 (HB), 9781108706711 (PB), 9781108588782 (OC)
ISSNs: 2517-7273 (online), 2517-7265 (print)

Contents

1 Introduction 1

2 The Social Dimensions of Scientific Knowledge 1

3 A Conception of Individual and Collective Scientific Knowledge 10

4 Assessing a Scientific Consensus for Knowledge and Dealing with Dissent 32

5 The Interaction of Epistemic and Social Elements in the Coproduction of Knowledge 49

References 72

1 Introduction

This Element is about the social dimensions of scientific knowledge. Section 2 asks in what ways scientific knowledge is social. Section 3 develops a conception[1] of scientific knowledge that accommodates the insights of the first section and is consonant with mainstream thinking about knowledge in analytic epistemology. Section 4 asks under what conditions we can tell, in the real world, that a consensus in a scientific community amounts to shared scientific knowledge, as characterized in the second section, and how to deal with scientific dissent. Section 5 reviews the ways in which epistemic and social elements interact to coproduce scientific knowledge.

This Element engages with literature from philosophy of science and social epistemology, especially social epistemology of science, as well as Science, Technology, and Society (STS) and analytic epistemology. These disparate scholarly traditions have hardly engaged with each other, and this Element strives to bring them into interaction. I recommend that nonphilosophers, especially from STS, read the sections in reverse order. The Element focuses on themes and debates that date from the start of the second millennium,[2] and on nonformal approaches to social epistemology.[3]

2 The Social Dimensions of Scientific Knowledge

Science is a social enterprise. Researchers collaborate, socially interact with each other, divide scientific tasks among them, and form social structures. While a primary aim of science is generating scientific knowledge, rarely do we find a developed conception of *scientific knowledge*, particularly one that takes seriously its social dimensions. Section 3 proposes such a conception. This section discusses the desiderata from it.

---

[1] While the view that knowledge is a social phenomenon is prevalent in philosophy of science, feminist, and social epistemology, it's underappreciated in orthodox analytic epistemology (McKenna 2022). An argument against it is that "infants and animals have knowledge ... thus the nature of knowledge cannot depend on language use, social needs, or distinctively human values" (Gardiner 2025). But as Gardiner argues, the aim of a social *conception* of knowledge (such as the one I develop in this Element) is to *ascribe* knowledge; namely, to correctly apply the concept of knowledge to cases. Whether knowledge is a social phenomenon or not, "knowledge ascriptions function to meet human needs, and the fundamental features of knowledge ascriptions stem from their roles in human social life. Ascription behaviour is infused with human interests and values" (Gardiner 2025). Additionally, animal knowledge, which exceeds the scope of this Element, also has social dimensions.

[2] Earlier debates about the social dimensions of scientific knowledge focused on patterns of rational theory change; see Lakatos and Musgrave (1970) and Nickles (2021).

[3] For computational social epistemology, see O'Connor (2023).

2.1 Scientific Knowledge: Propositional and True?

Knowledge is recognized in philosophy of science as a central aim of science (Bird 2022). A celebrated tradition in philosophy of science debates the conditions for the "growth of knowledge" (Lakatos and Musgrave 1970), yet one rarely finds a developed account in philosophy of science of that very thing the growth of which philosophers of science debate. While philosophers of science have developed detailed accounts of scientific explanation, scientific understanding, and scientific laws, they have largely neglected scientific knowledge.[4]

Philosophers of science, so it seems, have been content to leave the analysis of knowledge to analytic epistemologists. According to standard accounts of knowledge in analytic epistemology, an epistemic subject S knows that p only if

(1) S believes that p;
(2) p is true;
(3) p is justified.

The subject S is a single human person, and p is a proposition (which is, roughly, the content of a factual claim). For example, Jacob knows that there is an apple on the table only if Jacob believes that there is an apple on the table, his belief is justified; namely, he has visual evidence that supports this belief or his belief was reliably generated by his cognitive system, and his belief is true; namely, there is an apple on the table, rather than, say, a realistic decorative apple made of wax. Standard analyses typically add bridging conditions between (2) and (3), or substitute (3) with a modal (counterfactual) truth-tracking condition or a condition that credits the truth of S's belief to S's epistemic virtue.[5]

Leaving the analysis of knowledge to analytic epistemology, however, may have been a mistake. Standard analytic accounts of knowledge don't smoothly square with how philosophers of science, scientists, science teachers, science students, and the public often think about *scientific* knowledge. For philosophers of science, rather than justified or truth-tracking true belief, scientific knowledge typically means something like "the descriptive content of our best [scientific] theories and models and the skills and concepts required to understand this content" (Chakravartty 2022, 5).

What challenges does scientific knowledge pose to the traditional analyses of knowledge? Regarding the first condition, philosophers of science (notwithstanding Bayesians) are less concerned with a scientist's personal beliefs, and more with what that scientist publicly accepts for the sake of research. As Kent Staley and Aaron Cobb (2011), for example, write:

---

[4] Notable exceptions are Longino (2002), Roush (2005), and Suppe (1993).
[5] See further discussion in Section 3.4.1.

> While beliefs are certainly relevant to actions particular scientists perform, including the activity of endorsing particular experimental conclusions, the assertion and endorsement of such conclusions in these forums can be distinguished from an individual scientist's beliefs about these assertions. Although not conclusive, these considerations suggest pursuing the idea that scientific knowledge should not be understood as essentially a species of belief. At any rate, whether or not this is the case, one seeking to understand epistemic justification in the sciences and the practices that produce it would be better off looking to what is asserted in the appropriate social contexts than worrying about underlying beliefs, as it is through the interaction of communicative acts that the corpus of scientific knowledge is formed. (p. 479)

Additionally, scientific knowledge isn't necessarily propositional. "Traditional epistemology formulates its problems and answers by thinking of knowledge as primarily propositional. This presupposition should be scrutinized in the light of historical and sociological analyses of cognitive performance and in the light of contemporary theories of human cognition" (Kitcher 1992, 80–81). On top of propositions, scientific knowledge consists of abstract models, pictures, diagrams, graphs (Perini 2012), and even material models (Baird 2004, ch. 2).[6] Sometimes the information contained in them cannot be fully expressed in verbal propositional form, and even when it can, scientists don't bother to do so (Perini 2005). It might be objected that such representations don't constitute knowledge, but only propositions about them. But that isn't how the concept of knowledge is commonly used. When a neat philosophical theory doesn't square with messy reality, we shouldn't reject reality; rather, we should revise the theory.

The nonpropositional form of some scientific knowledge bears on the truth condition. Truth is a property or state of propositions, not of models, diagrams, and such. There are "multiple forms of semantic success, including truth, but also isomorphism, similarity, approximation, and others" (Longino 2022, 176). Unlike truth, which is binary, other semantic relations of fit are gradational, and their adequacy is determined in light of the purposes of the representation (Frigg and Hartmann 2020; Parker 2020).

Another difficulty with truth is that scientific knowledge is tentative: revisable and changeable. Yet scientists aren't quick to deny some past theories the status of knowledge although these theories have been refuted or revised. Scientists similarly don't deny present empirically successful theories the status of knowledge although these theories may be revised or refuted. Newtonian mechanics, for example, is, strictly speaking, a false theory, yet it's still taught

---

[6] For an overview of these differences, see Kitcher (1992; 1994), Longino (2002, ch. 5), Giere (1988, ch. 3), and Gerken (2019).

and used, since it's a good enough approximation for a wide variety of purposes. It's still part and parcel of the corpus of scientific knowledge.

A further difficulty with the truth condition concerns unobservable entities and processes. Much of scientific knowledge is about unobservable entities, such as protons, or about slow, large-scale processes, such as evolution, which cannot be directly perceived by a human observer. Antirealist philosophers are skeptical or agnostic about the truth of scientific claims about unobservables. Even realists don't think that *all* scientific knowledge about unobservables is true, but only a subset of scientific claims about unobservable entities. Often, they talk about "approximate truth" rather than truth simpliciter (Chakravartty 2017).

More radical worries about truth exist. Some argue that truth, understood as a mind-independent relation of correspondence between a representation and its target system, is irrelevant or unhelpful to the analysis of scientific knowledge. I will address these worries in Sections 3.5 and 3.6.

2.2 Scientific Knowledge: Distributed and Communal

Traditional analyses of knowledge are individualistic. They describe an isolated individual subject without mentioning their social standing. They regard facts about the subject's social standing as irrelevant to the analysis of knowledge. By contrast, social epistemologists regard facts about the subject's membership in an epistemic community; the evidence available within the community; the norms of inquiry, reasoning, and testimony that prevail in that community; and perceptions of risk that stems from error, as having substantive epistemic relevance to any account of scientific knowledge.

A common view in the social epistemology of science is that in some substantive sense, scientific knowledge is communal, and in some respects, communal knowledge is more fundamental than individual knowledge. Namely, an individual's epistemic dependence on members of their epistemic community goes beyond mere reliance on them as informants. Rather, whether that individual's personal beliefs count as knowledge depends on substantive constituents of knowledge that aren't or never have been theirs but are located in other members of their epistemic community or the community itself. An account of scientific knowledge should accommodate the following thesis.

> *The Epistemic Dependence Thesis*
> Whether a subject S knows p at time t may depend on epistemically substantive elements that other member(s) of S's epistemic community or the community itself possess at t, or an epistemically relevant event(s) undergone by some of them, individually or collectively, prior to t.

There are different versions of the *Epistemic Dependence Thesis*. The substantive epistemic elements that other members of the community or the community itself possess may be background beliefs (Nelson 1993; Longino 2002), evidence (Hardwig 1985; Miller 2015), shared epistemic standards, shared assessments of the severity of inductive risks (risks that stem from making a wrong epistemic judgment) (Wilholt 2009), or shared models of supportive reasoning (Kuhn 1970; Kusch 2002). Epistemically relevant events undergone by other members of the epistemic community may be belief-generating cognitive processes (Goldberg 2010), exercises of epistemic virtues (Green 2016), or critical scrutiny (Longino 2022). Such events aren't merely prior, necessary enabling conditions for the existence of knowledge, but substantive epistemic conditions that need to be met for scientific knowledge to obtain.

2.3 The Arguments for the Epistemic Dependence Thesis

The Argument from Distributed Cognitive Labor: cognitive processes that generate a single subject's beliefs and confer justification or the status of knowledge on these beliefs may occur largely outside the subject's own cognitive system and be distributed among multiple human subjects that epistemically depend on each other; scientific knowledge cannot be produced without such distribution of cognitive labor (Hutchins 1995; Giere 2002; Longino 2002, ch. 4; Goldberg 2010). Some versions of this argument extend cognition to technological artefacts as well as human subjects (Giere 2003; 2012; Heersmink 2016).

The Argument from Extended Justification: whether a subject's beliefs are sufficiently justified to constitute knowledge may depend on justificatory elements (e.g., evidence, segments of reliable belief-forming processes, cognitive virtues) of other members of their epistemic community whom the subject must explicitly or implicitly trust (Hardwig 1985; Faulkner 2018; Goldberg 2010; 2021; Miller 2015; Miller and Freiman 2020; Pritchard 2010; Green 2016; Dragos 2021; de Ridder 2014; 2019).

The Argument from the Sheer Vastness of Evidence: for some scientific claims, the evidence or arguments that are needed to justify them are so vast, that no individual on their own can review them; different individuals review small parts of evidence, and the substantive evidence by virtue of which such claims constitute knowledge is located only within a social collective (Habgood-Coote and Tanswell 2023; Hardwig 1991; Kukla 2012).

The Argument from Underdetermination of Theory by Evidence: more than one theory can fit the same evidence. Communal background assumptions, which aren't any particular individual's beliefs, stabilize individuals' justified beliefs in, or acceptance of a particular theory among the various possible theories. Additionally, when more than one available theory can accommodate the available data, shared communal values determine which one is accepted and certified as knowledge (Barnes and Bloor 1982; Kuhn 1970; Nelson 1993; Longino 2002; 2016; Potter 1996; compare Section 5.3).

The Argument from the Transformative Role of Criticism: typically, on scientific matters, even a highly reflective, open-minded, individual subject cannot free themself of all their biases and reach knowledge-level justification for their beliefs or claims on their own. To reach knowledge-level justification, hypotheses must undergo communal critical exchange that weeds out individuals' biases and examines a full range of alternative interpretations of the same evidence (Longino 2002, ch. 5; Nagel 1979, 13–14).

The Argument from Shared Reasoning Patterns: reasoning patterns, which must be followed for reasoning conclusions to constitute knowledge, are communal model solutions to cognitive problems (exemplars); for reasoning conclusions to constitute knowledge, the community must also be satisfied that these patterns were correctly followed (Kuhn 1970; Kusch 2002).

The Argument from Conventional Reliability Standards: epistemic reliability standards, which claims must pass to constitute knowledge, are communal conventions that reflect communal rather than individual weighings of inductive risks; without them, scientists cannot assess their peers' reliability and share their findings with each other (Wilholt 2009).

The Argument from Knowledge as a Common Good: some scientific knowledge is an inherently common good, produced and consumed collaboratively by many (Radder 2017).

The Argument from Theory-Ladenness of Observation: individuals' empirical observational beliefs are shaped by and acquire their meaning only through the mitigation of prior socially acquired and accepted theories and observational skills (Kuhn 1970).

2.4 Value-Ladenness and Pragmatic Encroachment

A second thesis this Element endorses is that *scientific knowledge is laden with social values*. Heather Douglas (2000; 2009, ch. 5) has influentially argued that science, specifically, the context of justification,[7] is neither free of social values nor should be.

---

[7] In the context of discovery, hypotheses are raised and research priorities are set, while in the context of justification, hypotheses are validated. It's less controversial that social values may legitimately influence the context of *discovery* by triggering hypotheses or setting research priorities. It's also acknowledged, however, that scientists should have autonomy to set research agendas. For science–society relations in the context of discovery, see Kitcher (2001; 2011), Keren (2013; 2015), and Jasanoff (2004c).

Douglas distinguishes two roles of social values in epistemic judgments: *direct* and *indirect*. In their direct role, values serve as reasons for making an epistemic judgment, such as theory acceptance. For Douglas, the direct role isn't legitimate in the context of justification, as it amounts to wishful thinking. In their indirect role, values determine the threshold level that evidence must meet for making a justified epistemic judgment by determining the levels of inductive risks we are willing to tolerate. Douglas identifies two types of inductive risks: wrongly accepting a false hypothesis, and wrongly rejecting a true hypothesis. There is an inherent trade-off between them. The more we expose ourselves to one type of risk, the less we expose ourselves to the other type.

Douglas argues that the indirect role is legitimate and required, because social values determine acceptable risks in a given context, and different social circumstances may legitimately require different balances between types of errors. When we value the possible consequences of a risk as mild, we lower the threshold level of evidence required for making a justified judgment, and when the consequences are acute, we raise it. (Section 5.4 reviews other ways values adjust evidential weights.)

Douglas (2009, 50–55) draws on Richard Rudner's (1953) paper "The Scientist *qua* Scientist Makes Value Judgments." The words "*qua* scientist" are important. Nobody denies that in some contexts, scientists make value judgments, for example, when they vote *qua* citizens. Rudner's point is that scientists make value judgments as an integral part of normal scientific practice within the context of justification. Rudner's argument has met criticism. One reply to Rudner is that scientists can explain inductive risks to relevant decision makers who decide which theory to apply. Even when such decision makers are scientists, they don't make them *qua* scientists, but *qua* politicians, civil servants, executives, and so on (Gundersen 2018).

This reply assumes that research consequences evaluation can be deferred to the context of application when inquiry is finished; thus, scientists, who work in the context of justification, needn't evaluate social risks. Douglas argues that this reply is inadequate because a strict separation between basic and applied science, or the contexts of justification and application, is unsustainable. Values penetrate deep into the context of justification and affect various stages of research, such as study design, data analysis, evidence characterization, and evidence interpretation. Analyzing a case study involving scientific research of the carcinogenic effects of dioxin, Douglas illustrates the inevitability of making value judgments in various stages of inquiry prior to theory acceptance. Douglas discusses a series of studies that exposed rats to dioxin. Slides with the rats' liver tissues were observed and characterized to determine if they had developed cancer. Four different studies that used the same slides as data characterized some of them differently, which led to different results regarding the carcinogenicity of dioxin.

Douglas argues that this case illustrates the necessity and inevitability of the indirect role. Values shouldn't influence the characterization of clear evidence. Clear cases of diseased tissues should be characterized as such, and clear cases of healthy tissue should be characterized as such. But values ought to influence the characterization of borderline evidence. In a society mostly concerned with the dangers of cancer, borderline slide cases should be characterized as diseased, and in a society mostly concerned with the economic burden of over-regulation, they should be characterized as healthy. Such a practice reflects the types and levels of inductive risk that society is willing to take (Douglas 2000; 2009, 124).[8]

Douglas' claims about an indirect role for values in epistemic judgments in science correspond to, and support a view of *knowledge* known as Pragmatic Encroachment, the proponents of which also draw on Rudner's (1953) classic paper (Fantl and McGrath 2011). Pragmatic Encroachment states that facts about a subject's practical interests regarding a certain content are relevant to determining whether their belief or acceptance of this content passes the threshold of knowledge-level justification. Specifically, if the subject has high stakes regarding the content, they are *ceteris paribus* in a worse position to know it than if they have low or no stakes regarding it. The logical relations between knowledge, justification, and interests according to Pragmatic Encroachment are analogous to the relations that Douglas draws between an epistemic judgment, evidence, and social values in their indirect-role capacity (Miller 2014b).

It might be objected that there is an objective, invariant threshold of justification for knowledge, independent of pragmatic factors. Perhaps such a standard is certainty or near certainty. Surely, if in daylight conditions I clearly see a cup on my desk, I can be certain, hence know that there is a cup on my desk. Such self-contained simple examples of perceptual knowledge of macro objects, however, lack essential features of knowledge that relies on complex real-world inquiry and involves dependence on others (Code 1993). Scientific knowledge often lacks the certainty of knowledge of a cup on a desk, but it's knowledge nevertheless.

Conee and Feldman (2004) suggest that an invariant, nonpragmatic standard for knowledge-level justification would be "along the lines of the legal standard for conviction in criminal cases, proof beyond a reasonable doubt" (p. 296). They argue such a standard would strike the right balance between not being too lax or too stringent. It's unclear, however, why a standard that reflects the

---

[8] For the arguments concerning the indirect role for values in science, see Elliot (2022, 22–28).

weighing of inductive risks in a criminal trial (better exonerate a guilty person than convict an innocent one) would be suitable for all knowledge, including scientific knowledge. As Conee and Feldman (2004), write, their notion of proof is "weaker than a mathematical proof" (p. 296). So why would it do, say, for mathematical knowledge?

Similarly, Gregor Betz (2013) argues that values needn't set a threshold of justification because "there is a class of scientific statements which can be considered – *for all practical purposes* – as established beyond reasonable doubt" (p. 218; emphasis added). But the caveat "for all practical purposes" explicitly states that practical concerns *are* relevant to justified theory acceptance. Betz simply weighs values such that they set the threshold for knowledge very high. Betz only shows that a class of scientific hypotheses trivially satisfies the practical conditions for their justified acceptance; namely, the risks associated with their justified acceptance in conceivable contexts are negligible. It doesn't follow that practical conditions are irrelevant for making justified scientific epistemic judgments (Miller 2014b).

Gerken (2019, 125–127) accepts that pragmatic factors are relevant to theory acceptance but argues that scientists ascribe *knowledge* to each other based on evidence alone. Gerken's objection fails, however. First, the conception of scientific knowledge developed here includes, for good reasons, acceptance and not just belief. But even if we restrict scientific knowledge to belief, knowledge ascription requires determining whether the evidence passes a threshold of knowledge-level justification. As argued, it's hard to see on what nonarbitrary, nonpragmatic basis a threshold that will cover all instances and uses of scientific knowledge can be formulated. Third, on complex scientific matters, such as global climate change, there is no candidate for *belief* other than best *accepted* scientific theories, which are saturated with the value judgments and trade-offs that were made in the inquiry process leading to their acceptance (Miller 2014b).

As I argue in Section 5.4, pragmatic factors can lower and raise evidential thresholds only within a limited range. Thus, while Gerken (2019, 126) is right that it would be weird, for example, to ascribe to a researcher knowledge that molecules X and Y bind *only because* nobody cares (assuming that the evidence is flimsy), when the evidence is stronger, pragmatic factors may make a difference in whether some content amounts to knowledge or not.

Only social values offer a nonarbitrary, principled, and relevant basis to decide the various dilemmas that arise during inquiry and influence its outcomes. Proponents of the Value-Free Ideal have yet to provide a persuasive argument for how one can set an evidential threshold for a justified epistemic judgment without appealing to social values, or how one can defer all these judgments to outside/after the normal process of scientific research. While the value-ladenness of scientific knowledge and Pragmatic Encroachment aren't undisputed, the conception of knowledge that the next section develops aims to accommodate them.

3 A Conception of Individual and Collective Scientific Knowledge

In the previous section, I argued that philosophy of science has largely neglected the analysis of knowledge, while analytic epistemology has developed analyses of knowledge that don't square well with essential features of scientific knowledge. In this section, I develop a conception of scientific knowledge that accommodates the special features of scientific knowledge (Section 2.1), the *Epistemic Dependence Thesis* (Section 2.2), and the value-ladenness of scientific knowledge (Section 2.4).

My conception is "backward compatible" with analytic accounts of knowledge in two senses. First, it doesn't deny that true propositional belief that satisfies certain conditions may be knowledge.[9] Rather, it lets in other forms of cognitive commitment on top of belief, nonpropositional content on top of propositions, and semantic relations of fit on top of truth. Second, it preserves and explains prevailing philosophical intuitions on standard examples in the analytic literature of subjects' having or lacking knowledge.

3.1 Agential versus Nonagential Accounts of Collective Knowledge

Accounts of collective knowledge are dividable into *agential* and *nonagential* (see Miller 2015, 420). Agential accounts assume that to have collective knowledge, members of a group must be able to act together as a body: to make collective decisions, set collective goals, make collective statements, and so on. They assume there is a corporate agent such as a committee or a government agency that holds a belief, and its being justified means that this agent should properly respond to evidence similarly to how an individual agent would respond to it (Lackey 2016; Silva 2019; Tollefsen 2015, ch. 1).

By contrast, nonagential accounts (Kusch 2002; Nelson 1993; Hardwig 1985; de Ridder 2014; Miller 2015) don't assume that. They view knowledge as collective in that it's a participatory good, produced and used by many, and stored in shared repositories. Knowledge is collective like language: just like a living language needs a linguistic community, scientific knowledge requires an epistemic community. But a linguistic community isn't an agent. Nonagential accounts ask whether putative knowledge meets, qua content, the evidential standards of the community, and whether these standards are adequate.

---

[9] But see Section 3.4.

There seems to be some confusion between agential and nonagential accounts of *scientific* knowledge. When some accounts ask what types of collective agents can have scientific knowledge (Bird 2022, ch. 4; Dragos 2016), they seem to actually ask what kinds of scientific groups are structured and organized such that they produce collective knowledge in the nonagential sense, rather than what scientific groups act as epistemic agents that navigate their way in the world like individual epistemic agents do.

Longino's and my proposed conceptions of collective knowledge are nonagential. My conception doesn't involve group belief, though it's compatible with it. Under my conception, to constitute knowledge, the belief of a corporate body, such as a committee or a government agency, should meet the same conditions as an individual's belief, including believing responsibly. Only agential knowledge, as opposed to nonagential knowledge, can form a basis for an agent's justified action, whether it's an individual agent or a corporate agent.

3.2 Longino's Account of Collective and Individual Knowledge

Longino's (2002, 128–140) pioneering analysis of scientific knowledge is almost the only existing account of scientific knowledge that is both in the spirit of the traditional analyses of knowledge and takes these considerations seriously. Longino's account of *communal* knowledge is this:

> A given content, A, accepted by members of [a community] C counts as *knowledge* for C if A conforms to its intended object(s) (sufficiently to enable members of C to carry out their projects regarding that/those object(s)) and A is epistemically acceptable in C. (Longino 2002, 136)

Longino's analysis is sensitive to the special features of scientific knowledge that were discussed in the previous section. To square with the fact that scientific knowledge isn't necessarily propositional, Longino talks about "content." Since the notion of truth simpliciter is problematic for characterizing scientific knowledge, she talks about conformation, which is a family of semantic relations between a representation and its target system suitable for the achievement of certain practical goals. Instead of justification, she talks about "epistemic acceptability," which is the epistemic standing of the content in an epistemic community.

Epistemic acceptability (communal justification) is at the center of Longino's account. Longino argues that when researchers theorize from empirical data, they may make biased background assumptions that taint their resulting theories. Since more than one theory can accommodate the same data, and the same data can stand in different evidential relations to the theory in light of different theoretical background assumptions, the theoretical background assumptions that scientists make in order to reason from evidence to theory should be as free from bias as possible (see Nagel 1979, 13–14).

Longino prescribes four social-epistemic norms of critical scrutiny to weed out such biases: (1) public venues for criticism; (2) criticism is taken up by the community; (3) the standards for criticism are public; (4) tempered equality of intellectual authority: intellectual capacity is the only criterion for participating in a knowledge-generating discussion, regardless of gender, race, and so on.[10] Critical discussion governed by these norms is *required* for theories to reach *knowledge-level justification*. Accordingly, Longino (2002) defines the epistemic acceptability condition as follows:

> Some content A is *epistemically acceptable* in community C at time *t* if A is or is supported by data d evident to C at *t* in light of reasoning and background assumptions which have survived critical scrutiny from as many perspectives as are available to C at *t*, and C is characterized by venues for criticism, uptake of criticism, public standards [for criticism], and tempered equality of intellectual authority. (p. 135)

For Longino, individual knowledge is derivative from communal knowledge. A subject knows if they can respond to critical challenges according to their community's justification standards. Longino's (2002) conditions for *individual* knowledge are as follows:

[An individual subject] S *knows* that *p* if
  i. S accepts *p*,
  ii. *p* (or *p* conforms to its intended object sufficiently …),
  iii. S's response to contextually appropriate criticism of *p* or of S's accepting *p* is or would be epistemically acceptable in C (i.e., S's response does or would satisfy standards adopted by C, and would itself be so evaluated by the relevant subgroup of C, in situations characterized by the conditions of effective criticism (p. 138).

There are, however, difficulties with Longino's account. Regarding communal knowledge, an epistemic community may follow Longino's epistemic acceptability procedures, but the accepted theory may still fall short of knowledge-level justification. This is because critical norms guarantee neither that community members actually raise and rebut relevant objections, nor that the members come up with, and rule out alternative theories that accommodate the same empirical data.[11] On the other hand, some proper and legitimate scientific knowledge has not been achieved through Longino's prescribed procedures (Solomon and Richardson 2005; Goldman 2002).

---

[10] See Miller and Pinto (2022) for an account of tempered equality.
[11] See more about this in Section 3.6.

Longino's account of individual knowledge is essentially the same as that of communal knowledge. The main difference is that the standards of criticism an individual should meet should be "contextually appropriate." For her, the epistemically significant interactions happen when communal social-epistemic norms solidify, rather than when already solidified norms affect the justificatory status of individuals' beliefs. Longino (2022) asks, "given that norms of reliability can be articulated independently of the subject's social context, how central are the social expectations to an individual's justifiedness?" (173).

Longino's account of individual knowledge, however, is problematic for two reasons. First, Longino doesn't sufficiently distinguish between different community members. Should an expert geologist and a layperson who accepts her testimony both be able to respond to the same critical challenges? Should all geologists be able to equally address all critical challenges regardless of their subspecialization? While to pass as knowledge, content should pass community-wide epistemic norms, to pass as knowers, different socially situated members of the community may face different legitimate social-epistemic expectations.

Second, justifying a theory is a joint process in which different members bring their own perspectives to criticize a theory *together*. An individual subject who claims individual knowledge typically lacks the resources, time, will, expertise, and cognitive ability to withstand alone the same effective criticism that the content that he claims to know should have passed to constitute communal knowledge (Hardwig 1985; Miller 2015). And why should he? After a theory has been communally certified as knowledge, a subject shouldn't be required to rejustify what is already known. Rather, he should be able to rely on the communal justification process that has already occurred. Namely, the burden of justification an individual subject needs to face to have knowledge, qua individual, typically should be lower than the burden of justification the scientific community should meet to have the same knowledge, qua community.

In sum, one may agree with Longino that individual knowledge is derivative of communal knowledge, but still desire an account of individual knowledge that would correctly discharge and allocate differently socially situated individuals epistemic responsibilities, or correctly adjudicate scenarios in which individuals, qua members of an epistemic communities, have or lack knowledge. In the next section, I propose such an account.

3.3 A Full Conception of Collective and Individual Scientific Knowledge

I will now propose an integrated conception of individual and collective scientific knowledge that builds on, but attempts to overcome the difficulties with

Longino's account. My proposed account is sensitive to the special features of scientific knowledge, the *Epistemic Dependence Thesis*, the value-ladenness of scientific knowledge, and the different epistemic standards individuals and communities typically need to meet, respectively, to have knowledge. For reasons that will be explained momentarily, I call this the *Full Conception of Scientific Knowledge*. It comes in two versions, individualistic and collective.

Scientific Knowledge (Full Conception, Individualistic)

An individual subject S knows a content p if and only if the following conditions are met:

Veracity:	p is true, approximately true, empirically adequate, there is appropriate semantic fit between p and its target system, and so on.
Commitment:	S believes, publicly accepts, endorses, or is otherwise committed to p for the sake of research.
Individual Responsibility:	S's belief that/acceptance of/commitment to p is epistemically responsible.
Collective Justification:	p is justified by S's epistemic community's evidential standards, and these standards employ evidential thresholds that correspond to a just and politically legitimate communal weighing of inductive risks (regardless of S's own evidence for p).
Objective Justification:	p is justified by the evidence the community jointly possesses in light of the justification requirements dictated by the objective situation (regardless of whether or not S or S's epistemic community are aware of the objective situation).

Scientific Knowledge (Full Conception, Collective)

A content p is knowledge in an epistemic community C if and only if the following three conditions obtain:

1. p meets the *Veracity* condition;
2. p meets the *Collective Justification* condition in C;
3. p meets the *Objective Justification* condition in C.

Like Longino's account, this conception of knowledge turns the traditional analyses on their head. While traditional analyses view collective knowledge as an aggregate of individual knowledge, my conception views collective knowledge as a shared repository of content that has passed community standards, and individual knowledge as what individuals gain when they successfully tap into this shared resource. I don't deny that individuals can have beliefs from their personal perception or inference, but I argue that for belief – at least in the scientific case – to gain knowledge status, it must satisfy certain community standards.

3.3.1 Veracity and Commitment

Let us review the account. *Veracity* is the truth condition relaxed to accommodate its difficulties with addressing the tentative or nonpropositional nature of scientific knowledge and skeptical worries about unobservable entities (Section 2.1). The *Veracity* condition is mind-independent and isn't relativized to an individual subject or an epistemic community. What determines whether a content that is subject to empirical testing meets the *Veracity* condition is the state of a world of extralinguistic things whose behavior is determined by natural laws or patterns that are independent of the rules and practices which determine how we speak and how we evaluate evidence.[12] I will defend this condition in Section 3.6.

The *Commitment* condition defines individual knowledge as belief, or acceptance or other public commitment. When I discuss belief, the discussion applies to acceptance too.

3.3.2 Individual Responsibility

Longino derives the conditions for *individual* knowledge from the conditions for collective knowledge but doesn't sufficiently distinguish between different individuals who are in different social positions and are required to meet different justification standards to have knowledge. The condition *Individual Responsibility*, which states that individually justified belief is individually responsible belief, addresses this difficulty, since responsibility standards for an individual epistemic subject vary according to her social position. This is a necessary condition for individual knowledge but not a sufficient one. For an individual's responsible belief to constitute knowledge, the content of the belief must still be collectively known. Namely, the other conditions need to be independently met.

Elsewhere Isaac Record and I (Miller and Record 2013) defend an account of individually justified belief as individually responsible belief, which is suitable for fulfilling the condition *Individual Responsibility*.[13] A belief is justified inasmuch as it's responsibly formed, where a responsible subject is one who does, inasmuch as they can, what is required of them in a particular situation to bring

---

[12] This formulation is a paraphrase on Bogen (1985, 196).
[13] See also Annis (1978), Foley (2005), and Kornblith (1983).

about true and rational beliefs. Individual Responsibility here is delimited in part by role-expectations within particular situations, while practicability is delimited by facts about the subject's competencies as well as their technological, ethical, and economic circumstances. Thus, according to our account, both what is practicable and what a subject is responsible to do will vary from person to person and from situation to situation.

Whether a belief is responsible depends inter alia on a subject's *stakes*. There is in principle no limit to how much evidence a subject may gather before forming a belief. Thus, the level of certainty a responsible subject is required to reach is partly determined by an appropriate weighing of inductive risks, namely, the practical consequences of error (Douglas 2000; 2009; see Section 2.4). For example, suppose a physician needs to form a belief about the best treatment for a patient with a rare disease. If she consults her old medical books, when she can easily access an online database that contains more up-to-date information, her belief is irresponsible. But if she doesn't have access to such a database, for example, she lives in the pre-Internet age, and her old books are the most reliable resource around, her belief may be responsible. If a high school student needs to make a small class presentation on this disease, and only her grade hinges on it, then to form a responsible belief on the best treatment, she needn't go through online databases.

Whether a subject's belief is responsible also partly depends on the *technological means of inquiry available to her*. By allowing or limiting subjects' attempts to gain knowledge, the availability or lack thereof of certain technologies effectively changes standards of epistemic responsibility and consequently justified belief.

Practicability provides both upper and lower limits for responsible belief. With respect to an upper limit, a responsible subject should obviously not be expected to do more than what is practicable. This doesn't mean, however, that a subject who has reached the upper limit, namely who has done all that is practicable to reach a justified belief, always achieves this aim. For example, the confirmation of a theory in particle physics may require a more powerful particle accelerator than the ones available.[14]

With respect to a lower limit, when a certain inquiry-related activity becomes practicable, performing it might become a minimal requirement for forming justified beliefs on certain matters. For example, in the early days of the COVID-19 pandemic, when swabs were scarce, a justified belief that a subject has contracted COVID-19 might be reached based on an examination of their symptoms and etiology alone. But when swab tests became widely available, performing them became a requirement for reaching a similar justified belief.

---

[14] See Miller and Record (2013, 126) for further analysis of different kinds of justification failures.

The Social Dimensions of Scientific Knowledge 17

The more practicable a putative justification-increasing investigative activity is, the likelier it is that it should be included among the subject's responsibilities (Miller and Record 2013, 125).

Unlike Longino's account of individual knowledge, *Individual Responsibility* doesn't require that a subject be able to defend a knowledge claim against critical challenges. In typical cases of seeking specialized scientific knowledge, to form a justified belief, a subject needs to locate a trustworthy expert or source and seek enough evidence to establish their credibility. A subject may critically evaluate some of the evidence the expert claims to rely on, but she needn't master the evidence in the community and be capable of withstanding cross-examination by other community members, as Longino seems to require.

3.3.3 Collective Justification

The *Collective Justification* condition expresses the principle that the justificatory status of a content (for an individual subject or for a scientific community) depends on the evidence in the community. It requires that the available evidence dispersed in the community support the subject's belief according to the community's epistemic standards in light of a normatively appropriate communal weighing of inductive risks.

Communal justification standards work similarly to individual epistemic responsibility standards. They are also affected by practicable means of inquiry and relevant inductive risks. In some social contexts, scientists may adopt lower evidential thresholds, while in others they may adopt higher ones (see Section 5.4).

The *Collective Justification* condition requires that the epistemic community employ evidential thresholds that correspond to a weighing of inductive risks that is both *just* and *politically legitimate*. The weighing of inductive risks takes into account all relevant interests of relevant stakeholders who may suffer from the harms caused by errors. For example, if the claim is that a certain substance that is used in the fast-food industry isn't harmful for human consumption, then the relevant stakeholders consist of consumers of fast food, and the relevant risk is harming their health. Which processes are politically legitimate and which distributions of risk are just greatly depend on the particularities of the case and the relevant stakeholders (see Section 5.4).[15]

With respect to political legitimacy, the knowledge-generation process in question may not be only the internal scientific process of research and validation, but also the external process of considering certain claims as a basis for public action. Science, Technology, and Society research shows that the stabilization and

---

[15] See Fernández Pinto and Hicks (2019) for an account of desirable values and roles for values in regulatory science, and Brown (2020) for roles for values in science in general.

acceptance or rejection of knowledge claims in policy contexts depends not only on their epistemic validity, narrowly construed, but also on the perceived political legitimacy of the process of their consideration. Which processes are regarded as politically legitimate greatly varies from one political culture to another (Jasanoff 2012). The requirement of political legitimacy turns this empirical insight into a normative requirement. It imposes restrictions on the process, such as a prohibition on massive lobbying, and adequate public and stakeholder representation (Brown 2009; Rolin 2009; Miller and Pinto 2022; see Section 5.5).

It might be objected that the fact that a weighing of inductive risks is unjust, for example, it unjustly imposes unfair risks on vulnerable groups, doesn't necessarily mean that the content in question isn't known, since knowledge is a matter of truth and evidence, rather than public ethics. As mentioned, however, scientific knowledge claims reflect the various trade-offs between values that were made in the process of inquiry leading to them, and the accepted truth-value of scientific knowledge claims is entangled with the value judgments that were made in the process of research that produced them. This means that if these value judgements are flawed, so are the knowledge claims. Thus, to constitute knowledge, it's required that the relevant value judgements not be flawed.

The *Collective Justification* condition, then, relativizes both individual and collective scientific knowledge to an epistemic community, specifically, its evidential standards, its weighing of inductive risks, and its practicable means of inquiry. Knowledge may differ in communities with different weighings of inductive risks even if they possess the same evidence. For example, California law ("Proposition 65") requires products that contain certain substances to bear a warning "this product can expose you to chemicals which are known to the State of California to cause cancer."[16] This warning seems puzzling. How can a substance be known in the State of California to cause cancer but not elsewhere? Does it cause cancer or doesn't it? It's certainly not meant to imply that the state of California has access to some secret evidence that others don't have.[17] My account easily explains this puzzle by appealing to California's weighing of inductive risks: California deems the error of labeling a noncarcinogenic substance as carcinogenic as less severe than other states or the federal government, and lowers the threshold of evidence to declare a substance carcinogenic accordingly.[18] It might be objected that this formulation is merely an artificial legal construct. This objection misses its target, however, because a legal construct needn't necessarily invoke the concept

---

[16] See www.p65warnings.ca.gov.   [17] I thank Torsten Wilholt for this observation.
[18] This case also involves a struggle for political authority (Section 5.5). Kukla (2010) argues that by labeling products as possibly carcinogenic, but not prohibiting them, California law shifts the blame for their potential harms from their manufacturers to their consumers.

of knowledge. When it does, we should still be able to explain what role the concept of knowledge plays in it.

Ceteris paribus, then, when are two communities that possess the same evidence for the same content in a situation such that this content constitutes knowledge for one community but not for the other? We should distinguish between two types of cases. In the first type of case, the two communities differ *only* in their assessment of inductive risks, and consequently in the way they respectively set evidential thresholds for accepting a knowledge claim. For example, suppose that California accepts a claim to knowledge according to which a substance is carcinogenic based on one well-conducted toxicity trial in rats, while the state of Montana requires at least two studies on two different animals. If the respective *Collective Justification* standards are set appropriately in both states – namely the weighing of risks is just (or at least morally defensible), achieved through a politically legitimate possess, and appropriately utilizes the available means of inquiry in both states – and the rest of the conditions for knowledge are met, then a substance for which there is only one positive well-conducted study in rats may be known in the state of California to cause cancer, but not in the state of Montana.

Things become more complicated when the two epistemic communities differ in epistemic standards other than their assessments of inductive risks or hold incompatible background assumptions, such as in Kuhnian incommensurability.[19] *In principle*, two such communities may both respectively legitimately deem two incompatible contents as collectively justified. *In practice*, however, such a situation should be extremely rare. For there to be two such epistemic communities, they should either have evolved in total isolation from each other, or, if they have been in contact, have experienced significant failure of communication. I don't think that this is the case for any human society. If they aren't totally isolated and don't suffer from severe failure of communication, critical cross-discussion, which would allow each society to critically examine its own epistemic practices by comparing them to the other's, can occur, and *should* occur if *Collective Justification* in each society is to be met.

It should be practically impossible to have two *modern scientific communities* that hold incompatible views, such that each satisfies the *Collective Justification* condition. This is because they both belong to the same global scientific culture.

---

[19] Kuhn (1970) famously argues that scientific research is organized around a "paradigm," which is, roughly, a theory and a set of epistemic and metaphysical commitments that accompany it. He argues that in some sense, scientists from different paradigms live in different realities, and there isn't an external perspective from which one reality can be said to be more correct than the other. Incommensurability between paradigms can be cognitive: scientists from different paradigms perceive reality differently. It can be semantic: they use different languages that don't converge on the same scientific entities. It can be epistemic: scientists use different standards for what counts as a legitimate scientific question and a satisfactory solution to it.

The researchers that participate in it share a commitment to establishing knowledge claims by empirical investigation. Researchers around the world have access to the Internet, which allows them to find each other, communicate, search databases for published papers, and in some cases also datasets. This means that scientists have the practicable technological possibility to know the state of research in their field of expertise, which means, according to the account of epistemic responsibility presented here, that they epistemically ought to do so (see Miller and Record 2013, 129–131). Even much before the Internet, science was a global phenomenon. Thus, a situation in which there are two *scientific* communities that have mutually incommensurable standards for collective justification and are both collectively justified is purely theoretical.[20] In other words, under my account, scientific knowledge doesn't come easily. Either two such epistemic communities have resolved the incommensurability to an extent they are both collectively justified in their respective theories, or they are both collectively unjustified.

To be clear: I don't assume that there is unity of scientific method. There is probably no one set of epistemic rules, procedures, methods, and standards that is shared by political economists, developmental psychologists, and particle physicists. The plurality of methods and modalities of evidence is both a strength and a weakness of modern science. When different lines of evidence produced by different methods mutually support each other, their conclusion is better supported, and more likely to achieve knowledge-level communal justification. When they don't, it typically doesn't reach collective knowledge-level justification. As Stegenga (2009) argues, exactly because different lines of evidence may speak different incommensurable languages and make incompatible background assumptions, determining whether multimodal lines of evidence support each other or not may itself be a difficult task.

I don't have a magic recipe for amalgamating multimodal evidence or rationally overcoming paradigm incommensurability. My point is merely that for the question of whether a modern scientific content satisfies the *Collective Justification* condition, the relevant community is the global scientific community, and the relevant evidence is the evidence at its disposal (cf. Goldberg 2021).

For the sake of satisfying *Collective Justification*, then, disciplinary or subdisciplinary boundaries don't shield scientists from evidence outside their discipline or subdiscipline. For example, in the first half of the nineteenth century, Avogadro's (1776–1856) (true) hypothesis that under the same conditions of temperature and pressure, equal volumes of all gases contain equal numbers of molecules, was accepted by some crystallographers, and rejected by nearly all

---

[20] I thank Ben Almassi for pressing me on this point. Goldberg (2018, 246–255) argues for a similar position.

organic chemists. From organic chemists' standpoint, although Avogadro had managed to establish the correct qualitative formulas for a large number of compounds, in many cases, his quantitative formulas were incorrect. By contrast, for crystallographers, who at the time had a better grasp of the distinction between "atom" and "molecule," the formula seemed accurate and useful. Only when the distinction between elements, atoms, and molecules became clear to all chemists, and it was theorized that some element molecules consist of more than a single atom of the same kind, was the hypothesis accepted (Lugg 1978, 282–283). Although the crystallographers had a better conceptual grasp of the relevant parts of chemistry and their views eventually prevailed, they did *not* possess collective justification until the organic chemists' challenges were met.

Similarly, nowadays, generative linguists and psycholinguists disagree about how children acquire language. Generative linguists think that children are born with a biologically innate universal grammar, which they learn how to use with a particular language, while psycholinguists find the idea of a biologically innate universal grammar implausible and think that children use general learning mechanisms to acquire language. Each theory has some empirical success (Lemetyinen 2012).[21] Since the two theories are mutually incompatible, from an epistemic perspective, neither group currently satisfies the *Collective Justification* condition.

Disciplinary divides, however, don't necessarily prevent the satisfaction of the *Collective Justification* condition. For example, Longino (2013) analyzes how a disciplinary differentiation of focus and of causal modeling occurs in scientific research of aggression and sexuality conducted by sociologists, psychologists, molecular geneticists, and neurobiologists. The different disciplines focus on different levels of description of the allegedly same phenomena, and on different coexisting but separate mechanisms that produce similar effects. Inasmuch as different disciplines that study the same phenomenon produce complementary perspectives rather than conflicting ones, then they *might* respectively satisfy *Collective Justification*. Yet, as Longino (2013, 204–207) observes, the knowledge that is produced may reveal how much isn't known about the phenomena in question. Longino further indicates that lack of sufficient interactions between the disciplines, which is manifested in few cross-disciplinary citations, is an epistemic deficiency. Such lack of cross-disciplinary interaction constitutes an impediment to satisfying the *Collective Justification* condition.

I stress again that *Collective Justification*, which applies to an epistemic community, is independent of *Individual Responsibility*. For having knowledge, qua an individual, a subject needn't necessarily be aware of all the

---

[21] I thank Roey Gafter for this example.

evidence in the community. If an individual responsibly believes a theory that enjoys collective justification (whether she's aware of that or not), then she also knows; if she responsibly believes a theory that doesn't enjoy collective justification, the subject has a justified belief, but not knowledge. Conversely, a subject may fail to responsibly believe a theory that enjoys collective justification. In such a case, the theory would be collectively known (if *Veracity* and *Objective Justification* are met), but the subject would not individually know it.

3.4 Can an Individual Know Irrespective of His Community?[22]

My account resembles Frederick Suppe's (1993) two-tier, individual and communal, account of scientific knowledge. For Suppe, an individual subject's true belief is knowledge if and only if it's reliably caused in a right way.[23] Causes for belief include the subject's personal observation and another person's testimony. For an individual researcher's knowledge claim to be admitted into the body of *shared scientific knowledge*, it must further undergo a *credentialing* process: it "must be augmented by sufficient justification to preempt or answer any specific doubts legitimized by that discipline" (1993, 163). Suppe's credentialing requirement is equivalent to my *Collective Justification* condition.

As opposed to my account, Suppe's account allows for a subject to have individual knowledge from personal observation even if it doesn't meet *Collective Justification*. Suppe (1993, 167–170) describes Lacy, a maverick scientist who reliably conducts an experiment but doesn't write or publish a research paper with the supportive argumentation for its result, namely, without credentialing the result. Suppe argues that Lacy can still personally come to know the result if his causal belief-forming process is truth-tracking in the right way.

I think that Suppe is wrong to think of communal credentialing as a process that transforms mere personal knowledge into somehow better, more justified credentialed scientific knowledge. The very point of knowledge as a cognitive achievement is that it requires no further upgrading. But I can see why this view could be attractive for someone who thinks that Galileo, for example, had knowledge although he didn't manage to get his views accepted by his peers.[24]

Note, however, that such merely personal knowledge is restricted in range and plays a limited role in science. *First*, direct personal observational knowledge, as opposed to testimonial knowledge, typically constitutes a tiny part of a researcher's overall knowledge (Hardwig 1985; 1991). *Second*, Lacy doesn't

---

[22] Miller (forthcoming) contains an extended version of the argument in this section.
[23] The technical details of what "in a right way" means are less important for this discussion.
[24] I thank Alex Bird for this objection.

necessarily know what he knows,[25] and without communal credentialing, cannot reliably self-attribute knowledge. Galileo, for example, held a mix of true and false beliefs about which he was equally confident, and between which we can distinguish only in retrospect. *Third*, Suppe acknowledges that for Lacy to *share* his personal observational knowledge with others through testimony, such that they also come to know, or to be recognized as a knower, Lacy must still have his knowledge claims credentialed by the relevant epistemic community.

Note that the third point follows from the second: since Lacy doesn't know what he knows, when it comes to his noncredentialed empirical reports, he doesn't reliably testify only when he knows, hence the audience of his noncredentialed testimonies doesn't satisfy Suppe's reliability condition, and doesn't gain knowledge from them. That humans are *not* generally reliable producers and consumers of testimony is a problem for any account that defines mere reliable generation of a true belief as a sufficient condition for knowledge (Miller 2015, 435–438; Shieber 2015, ch. 6).[26]

3.4.1 Objective Justification

The last condition is *Objective Justification*. The distinction between *Collective Justification* and *Objective Justification* is similar to Alvin Goldman's (1988) distinction between "weak" and "strong" justification: a belief formed by a process that is actually reliable is strongly justified, whereas a belief formed by a method that is considered reliable in an epistemic community is weakly justified and epistemically blameless. Adam (2007) and Stanley and Cobb (2011) make a similar distinction for scientific knowledge.

The condition of *Objective Justification* addresses the aforementioned difficulty with Longino's account of knowledge, which is that a community may deem justified a theory that fails to actually reach knowledge-level justification because it has not conceived and ruled out a relevant alternative hypothesis. An overlooked hypothesis is relevant in the sense that it could have relatively easily been the case that the hypothesis had met the *Veracity* condition.

For example, nineteenth-century physicists' (true) beliefs about cathode-ray tube phenomena before the discovery of X-rays were justified by communal standards, but not objectively justified, because scientists did not control for the possible influence of X-rays. As Kuhn (1970, 59) writes:

---

[25] Suppe (1993, 159) denies the KK-principle.
[26] Roush (2005) develops a truth-tracking theory of scientific knowledge that doesn't shoulder truth-tracking merely on the believing individual at the moment of belief formation. For Roush, subjects gain scientific knowledge only when science is truth-tracking over time. Roush's theory might be compatible with the desiderata in Section 1.

> Though X-rays were not prohibited by established theory, they violated deeply entrenched expectations [...] By the 1890s cathode ray equipment was widely deployed in numerous European laboratories. If Roentgen's apparatus had produced X-rays, then a number of other experimentalists must for some time have been producing those rays without knowing it. Perhaps those rays, which might well have other unacknowledged sources too, were implicated in behavior previously explained without reference to them.

Other examples of content that is collectively but not objectively justified include an experiment or a survey that is performed according to accepted methodological standards, but its participants try to manipulate its outcome for reasons the researchers could not have easily anticipated in advance (Vines et al. 2013; Fronek and Briggs 2018), or in which unexpectedly and in a manner that is hard to detect, an analysis software uses different parameters from the ones the researchers have set (Bateman and Napier 2011). Even if the final results in such cases are true and satisfy accepted methodological standards, they don't constitute knowledge because they fail to achieve *Objective Justification* (see de Ridder 2014, 44).

The objective situation, then, sometimes requires stricter justification standards than those required by the epistemic community. This, I will now argue, is also what happens in Gettier (1963) cases. A typical Gettier case goes as follows: you see a sheep-looking animal on the top of a hill. You form the belief that there is a sheep on the top of the hill. It's, in fact, a hairy, white, sheep-looking dog. But there is a sheep standing right behind it, which you don't see. Your belief is justified, true, but not knowledge (Chisholm 1977, 105).

Such cases – call them "first-generation Gettier cases" – seem suspect because they trade on an ambiguity in reference (the animal to which the true proposition "there is a sheep on the top of the hill" refers isn't the same animal you are thinking about) or on a formally allowed yet fishy logical inference, as in Gettier's (1963) original examples. A case that avoids such problems was introduced by Goldman (1976): you drive in the countryside and look at a barn. You form a true belief that it's a barn. Unbeknownst to you, you are looking at the only real barn in fake-barn country. Because you could have easily looked at a fake barn and believed it was a real barn, you don't know that it's a barn, although your belief that it's a barn is true. Unlike first-generation Gettier, there is nothing fishy going on here: you have competently and correctly recognized a barn as a barn. By chance, you were looking at the only real barn around. Call such cases and variations thereupon "second-generation Gettier cases" or "fake-barns cases."

In Gettier and fake-barns cases, it seems from the believing subject's perspective that they know, when, in fact, they don't, although their belief is true and from their perspective justified. As I will argue, the distinction I have drawn between collective and objective justification explains why this seems so to the subject.

Foley (2012) argues that what all Gettier and fake-barn cases have in common is that there is relevant information the believing subject lacks, which, if they had, they would not have formed the belief. Information gathering is a practical matter, thus whether a subject or a community has gathered or can gather this information partly depends on their stakes and available resources.

I add to Foley's observation that this missing information corresponds to evidence that is *exactly* in the gap between the evidence that the accepted communal justification standards require having in such a case, and evidence the particular situation objectively requires. Namely, in such cases, unbeknownst to the subjects, certain evidence is objectively required for ruling out a relevant alternative, which the communal standards of justification don't require to rule out; for example, that the structure is a fake barn rather than a real one. Namely, unlike the received view in analytic epistemology, which analyzes Gettier and fake-barn failures *knowledge* failures, I analyze such failures as *justification* failures.[27]

3.4.2 Implications of the Distinction between Individual, Collective, and Objective Justification

The distinction between collective and objective justification has implications to the analysis of epistemic scientific failures. Let's call cases of true claims that lack sufficient evidential support "under-substantiated positives." One kind of cases is when a scientist or a group of scientists knowingly fails to meet *Collective Justification* standards and is therefore epistemically irresponsible and blameworthy. Due to publication pressure, priority race, or overconfidence, scientists sometimes make shortcuts, and make discovery or confirmation reports not sufficiently backed by their evidence. When such reports turn out false, for example, Fleischmann and Pons' 1989 announcement of cold fusion, the researchers are penalized by the community. But when the reports are accepted as true, the scientists are usually praised and rewarded. Eminent scientists, including Eddington, Millikan, and Pasteur, have arguably made true discovery claims, for which they were hailed, although they were not sufficiently supported by the evidence at their disposal according to their respective community's standards (Miller 2015, 424–428).

---

[27] For analyses of Gettier failures as justification failures (rather than knowledge failures) see Scheman (1995, 188) and Bogen (1985, 202–203).

A *second* kind of cases of undersubstantiated positives, which has a parallel structure to Gettier cases, is when a scientist or a group of scientists meet *Collective Justification* standards but not the *Objective Justification* standards, and are therefore epistemically blameless qua individuals or qua a group (though it may still be asked whether the community standards are correct). The cathode-ray experiment example presented earlier corresponds to this type. Unlike false negatives and false positives, which have been recognized in social epistemology of science as cases of interest, undersubstantiated positives have largely flown under the radar.[28]

More generally, the distinction between individual-level justification (the *Individual Responsibility* condition), communal-level justification (the *Collective Justification* condition), and *Objective Justification* has implications for the analysis of the role of social values in science, especially their role in scientific failures. As mentioned in Section 2.4, Douglas (2000; 2009) argues that we should weigh inductive risks and allow social values to play an indirect role in evidence assessments that leads to justified epistemic judgments. We should manage two types of errors against each other: false negatives and false positives. My analysis suggests that for gaining knowledge, such error management should be done at two separate levels: the individual scientist, and the scientific community. The scientist may or may not have acted responsibly and traded inductive risks adequately, and may or may meet the communal justification standards. The scientific community may or may not set communal justification standards that legitimately and appropriately trade off different inductive risks. The weighing of inductive risks for individually responsible belief and for communal scientific knowledge needn't be the same or be influenced by the same values. Moreover, we may assign epistemic praise or blame at different intermediate levels between an individual scientist and the entire scientific community, such as the level of research team, a subscientific community, and so on. Additionally, a scientific community may or may not set communal justification standards that correspond to the *Objective Justification* standards, and if it doesn't, it may or may not be epistemically blameworthy for that.

Another family of cases of interest in the history of philosophy consists of *skeptical scenarios*, such as being misled by a Cartesian evil demon or being a brain in a vat. In such cases, it's impossible *in principle* to meet the *Objective Justification* standards. In skeptical scenarios, no matter how much inquiry is done and how stringent the communal justification standards are, attaining knowledge is impossible. Strictly speaking, a brain in a vat has no epistemic

---

[28] An exception is Magnus (2018), who draws on such cases to challenge the value-free ideal.

community. "Community standards" in such a case refer to *our* epistemic community, by which we implicitly assess the entity's beliefs. This is why, typically, such an entity is described as believing it's a regular person. A skeptic is one who takes seriously skeptical scenarios as relevant alternatives that must be ruled out to reach knowledge-level justification. But because the *Objective Justification* standards cannot be met, the skeptic concludes that we don't have knowledge, but at most, justified belief.

This analysis of skeptical scenarios is relevant to scientific knowledge in two ways. First, it explains why skeptical hypotheses are outside the domain of science. In science, hypotheses are judged based on evidence, but skeptical hypotheses are so constructed that no evidence can rule them out. Second, as Slater et al. (2020) argue, certain forms of science denialism adopt the logic of skeptical scenarios. Most of us don't have direct, personal, non-testimonial evidence that HIV causes AIDS or that the climate is changing. A denialist raises the knowledge-level justification standards such that they must conclusively rule out the possibility that all the available testimonial evidence is misleading, for example, a result of a huge conspiracy. Because an individual person cannot conclusively rule out this possibility (cf. Hardwig 1985; Miller 2015), the science denialist claims we don't know.

3.5 The Restricted Conception of Scientific Knowledge

In the previous section, I introduced my conception of scientific knowledge, which is attentive to the unique features of scientific knowledge, the *Epistemic Dependence Thesis*, and the value-ladenness of scientific knowledge. My conception includes a distinction between two conditions that correspond to two levels of justification: *Collective Justification* and *Objective Justification*. I argued that both levels are required for explaining common cases in mainstream epistemology literature in which it seems from a subject's perspective that she knows, where, in fact, she doesn't. I have also pointed out that there are real cases of interest from technoscience that share a similar pattern.

In this section, I present another motivation for distinguishing between collective and objective justification, which is that this distinction allows us to identify and characterize a commonly used conception of scientific knowledge, which the traditional analyses of knowledge in analytic epistemology have mostly overlooked.

Call this conception the *Restricted Conception of Scientific Knowledge*. It regards as collective scientific knowledge content that passes the *Collective Justification* condition, without necessarily passing the *Veracity* condition or the *Objective Justification* condition. Regarding individual knowledge, under the Restricted Conception, an individual subject knows when she believes or accepts *responsibly* content that passes the justification standards of her epistemic community, regardless of whether it's true (veridical) or objectively justified.

> **Scientific Knowledge (Restricted Conception, Collective)**
> A given content *p* is knowledge in an epistemic community C if and only if *p* meets the condition *Collective Justification* in C.

> **Scientific Knowledge (Restricted Conception, Individualistic)**
> An individual subject S knows *p* if and only if the following is true: S meets the *Commitment* and *Individual Responsibility* conditions, and p meets the condition *Collective Justification* in S's epistemic community.

An overlooked fact in mainstream analytic epistemology is that when researchers, practitioners, and nonanalytic philosophers invoke the concept of scientific knowledge, they often have the *Restricted Conception* in mind or something like it. For example, when Robert McHenry, former Editor-In-Chief of Encyclopedia Britannica, was asked about racially biased false claims in previous editions of the encyclopedia, he replied as follows:

> At Britannica, we tried to remember to describe what we were doing as setting down the state of our current knowledge. *We certainly never claimed to be writing the truth* (in Glosserman and Hill 2010, 1:03:09–1:03:26; emphasis added).

McHenry clearly has in mind a concept of knowledge as only what the scientific community deems justified at a given time, which corresponds only to the *Collective Justification* condition.

Indeed, the view that scientific knowledge is tentative and therefore not necessarily true or entirely true is widely shared among scientists. A 2019 consensus report about the nature of science by the National Academies of Science, Engineering, and Medicine (2019) states "scientific knowledge is durable and mutable." It adds that "[t]he advent of new scientific knowledge that displaces or reframes previous knowledge shouldn't be interpreted as a weakness in science" (33). Both old theories and new ones that displace them are considered in this statement as proper scientific knowledge.

The received view of scientific knowledge in mainstream philosophy of science in the empiricist tradition also resonates – though often implicitly – with the *Restricted Conception of Scientific Knowledge*. Logical positivists, Hempel, Quine, Longino, Popper, Lakatos, and Kuhn have all rejected in some way or another a mind-independent truth condition and regarded scientific knowledge as highly justified content.[29]

I have distinguished between two conceptions of scientific knowledge: the *Full Conception*, which (or something like it) is employed in mainstream epistemology, and the *Restricted Conception*, which (or something like it) is employed in scientific practice and in philosophy of science, though not always explicitly. One issue on which these two conceptions differ is whether scientific knowledge must be true, in the sense of mind-independent semantic fit. Another issue on which they differ is whether the *Collective Justification* standards of a scientific community are sufficient for knowledge, or an additional condition is required. I address this question in the next subsection.

3.6 Objective Justification Again

After having presented the *Restricted Conception of Scientific Knowledge*, one may ask: why isn't the *Restricted Conception* Enough? Why are the *Objective Justification* and *Veracity* conditions needed? This section addresses this question.

I have argued that the *Veracity* and *Objective Justification* conditions are required for explaining failures to achieve knowledge that are commonly discussed in analytic epistemology. An opponent to the *Veracity* and *Objective Justification* conditions, however, might not be impressed by this claim. A common sentiment among philosophers of science is that Gettier cases and their like are a misguided and unhelpful way to study knowledge.[30] It's therefore not a pressing philosophical concern to explain the failures in them. The cases from science that have a parallel structure to Gettier cases can be explained by the conceptual tools afforded by the *Restricted Conception*, or so this objection goes.

It may indeed be possible to analyze such scientific epistemic failures using the conceptual tools of the *Restricted Conception* alone. In this section, I argue, however, that this approach requires one to accept unqualified (or at least, not sufficiently qualified) epistemic relativism or to assume a priori there is scientific progress. This price is too high; therefore, opting for the *Full Conception* is preferable.

First, an objective justification condition is required to avoid epistemic relativism. Without it, we can only judge views accepted in an epistemic community as wrong or unjustified from another community's perspective,

---

[29] For statements to this effect, see Laudan and Leplin (1991, 466); Popper (1972); Kuhn (2000, 115, 160); Hacking (1983, 112–113). Pragmatists hold similar views (Noddings 2015, 110).

[30] While this sentiment is common among philosophers of sciences, it's hard to find it in writing. For two explicit statements, see Bogen (1985, 188) and Stich (1990, 3). I thank Amir Horowitz for the latter reference.

but not to prefer one community's perspective over another (Kusch 2002, ch. 18). For many, such relativism is objectionable.

It might be objected that a collective justification condition can avoid relativism if it's sufficiently stringent. Longino (2002) holds that a condition of collective justification that requires "critical exchanges in which the full variety of perspectives are represented and they are treated as equally capable of generating significant challenges to theories" (158) would winnow out claims to knowledge that shouldn't pass the threshold of knowledge-level justification. In particular, communities that subscribe to creationism, tealeaf reading, or witch-hunting (all counterexamples that have been raised against her account) would fail to certify their belief systems as knowledge, hence her account of knowledge doesn't collapse into relativism (2002, 158–162).

Indeed, an *ideal or nearly ideal* rational epistemic community would reach knowledge-level justification this way. But comparing the performance of real-world epistemic communities to an ideal community introduces an objective justification condition through the back door: like *Objective Justification*, an ideal community is a fiction of perfection of which any real community falls short.

Regarding real-world, imperfect communities, the discussion of communities of creationists, tealeaf readers, or witch-hunters is a red herring. The relevant point is that in the real world, critical norms don't *guarantee* that community members conceive of all the relevant critical challenges of the available data. It may also just so happen than no community member is bright and imaginative enough to think of all relevant alternative explanations.

Thus, we can have two communities that follow the same critical norms and consider a certain theory T. In the first community, a member conceives of an alternative theory R, the community finds R superior to T and accepts R as knowledge. In the second community, no member conceives of R, and the community accepts T as knowledge. Both communities find T and R, respectively, as sufficiently conforming to their target systems. Suppose that T and R are mutually exclusive. It follows that under Longino's account of knowledge, or under the restricted conception more generally, both communities know T and R respectively. Hence, even a stringent collective justification condition doesn't avoid relativism. Making the epistemic community itself the final arbitrator whether its own deliberation process is critical enough to be knowledge-yielding opens up the possibility that two communities would certify conflicting theories as knowledge, which leads to epistemic relativism.

A way to avoid relativism without subscribing to the *Objective Justification* condition is to assume that science progresses toward the truth or toward better justification practices, and to set current theories and justification practices as the standard for evaluating past theories and beliefs. We shouldn't assume, however, that science necessarily progresses. Even if one thinks that progress is an ideal for science, actual science may not achieve it. Nondirectional patterns of change in received scientific knowledge are possible. Discarded theories may be resurrected, such as Paul Ehrlich's (1854–1915) theory of antibody formation (Silverstein 2016), or the theory of inheritance of acquired traits, which has made a comeback (Jablonka and Lamb 2014). Scientific fashions may change, and with them, so do accepted theories, without later ones being necessarily better (Crane 1969). Or scientific knowledge may inadvertently affect the reality it purports to describe and be in a constant feedback loop process of reflecting the changes it has itself induced (Hacking 1999, ch. 4; 2007).

It might still be objected that the *Veracity* and *Objective Justification* conditions employ "the god trick of seeing everything from nowhere" (Haraway 1988, 581). Namely, they take a perspective from which reality is fully known. Yet no human being can occupy this perspective. Every human being's perspective is partial and subjective. Presenting a claim as if it's objectively justified is a means of concealing the real perspective from which it's made, which is typically the perspective of those in power. Like the clergy, who illegitimately present their own will as the will of God, when those in power declare that a claim has passed an alleged objective and neutral test of justification and truth, they enforce their own will, and suppress further challenges by the less powerful (Code 1993).

As long as we remember that no human can occupy an objective position from which all truths are known, however, the *Objective Justification* condition can be seen as an expression of modesty. Humanity isn't the measure of knowledge. There may be truths we can believe, but never know because we cannot meet the justification standards required to know them. Appearances and our best inferences may be misleading and wrong. *Objective Justification* is a humbling reminder of these possibilities.

3.7 Conclusion

I suggested a conception of scientific knowledge that is sensitive to its social aspects. My departure point was Longino's pioneering analysis of scientific knowledge, which defines knowledge in terms of social-epistemic norms a successful knowledge generation process must meet. I noted two main difficulties with Longino's approach. First, it doesn't sufficiently distinguish individual from collective knowledge. Second, following good processes doesn't guarantee a successful outcome.

To overcome these difficulties, I proposed a conception of scientific knowledge that specifies separate conditions for individual and collective knowledge and focuses on properties of outcomes, namely, evidential, political, and inductive-risk standards content must pass to constitute scientific knowledge. According to this conception, knowledge is *individually responsible, collectively and objectively justified true(ish) scientific content*.

Within the conception, I introduced an *Objective-Justification* condition to cover cases in which, unbeknownst to the researchers, they have failed to rule out a relevant explanation of their data or to control for an unknown factor that might have influenced their data. This condition is required for ruling out cases in which researchers follow communally prescribed knowledge-generating processes (such as the ones Longino formulates), but still fail to reach knowledge, even if they believe they have. The *Objective-Justification* condition is the lesser evil compared to epistemic relativism or an a priori assumption of scientific progress.

My account unifies individual and collective knowledge. It explains the role of inductive risks at the level of individual and collective knowledge. It consistently explains Gettier cases, fake-barns cases, and classical skeptical scenarios as cases that meet *Collective Justification* but fail to meet *Objective Justification*. It explains a common way of thinking of scientific knowledge by scientists as merely collectively justified belief or acceptance, which is more restrictive than the full-blown philosophical account of knowledge.

4 Assessing a Scientific Consensus for Knowledge and Dealing with Dissent

The conception in the previous section tells us what scientific knowledge is. It doesn't tell us whether a given person or community possesses knowledge, or whether a given content constitutes knowledge. So how do we tell whether a given content is knowledge, given that we don't have a magic truth detector, reality is messy, and our perspective is limited and partial. Although we aren't in a position to conclusively determine whether the *Objective Justification* is met, how can we nevertheless utilize our understanding of the social dimensions of scientific knowledge to identify concrete instances of it? This section addresses this question.

4.1 The Problem of Legitimate Knowledge Attribution

The question is when does a given scientific content amount to knowledge, as analyzed in the previous section. Another way to ask this question is: when may we legitimately attribute knowledge to scientists?

Sometimes this problem is pretty easy. *Easy cases* typically consist of well-established, settled, textbook knowledge in scientific domains with a long, proven track record of success. For example, it should generally not be disputed that nuclear physicists possess genuine knowledge of physics. Stephen Turner (2001) calls possessors of such knowledge "Type I Experts" (130).

Peter Vickers (2023) similarly identifies a class of scientific claims as "future-proof" science. A claim is "future proof," namely not subject to future revision, when there is a large, international, and diverse consensus on it that consists of at least 95 percent of the scientific community; for example, the claim that the earth has an iron core. We should add two caveats to this notion, however. First, as Vickers repeatedly admits, scientists should be kept away from this notion. Once scientists explicitly try to satisfy it and reach a 95 percent consensus, they may use illegitimate means such as coercion or suppression, and the criterion for future-proof science would lose its force. Second, Vickers assumes that modern science will continue work in more or less the same way. If science loses its core features that make it epistemically successful, Vickers' criterion will lose its force. What these core features are brings us back to the old question of demarcation of science from nonscience.

Another easy case of knowledge attribution involves what Stegenga (2018, ch. 4) calls "magic bullets" in medicine; namely, medical interventions the efficacy of which is specific and clear. Examples of magic bullets include insulin treatment for type 1 diabetes, and penicillin treatment for previously lethal bacterial infections. We can legitimately claim to know that magic bullets are effective.

We should, however, note three important caveats. *First*, easy cases don't include cases of epistemic trespassing, in which experts make authoritative knowledge claims outside their domain of expertise (Ballantyne 2018). For example, in May 2020, Israel's National Security Council formed an expert advisory panel on Israel's exit strategy from the coronavirus crisis. The (all-male) panel predominantly consisted of physicists, who had not dealt with health issues in their previous research (Prince-Gibson 2020). A straightforward attribution of coronavirus-related knowledge to this panel will be a mistake, even if its members' track record of success in physics is impeccable. While crossing disciplinary boundaries may be epistemically beneficial, cases involving such crossings aren't easy cases for the purpose of legitimate knowledge attribution.

Second, easy cases are restricted to well-established areas of science. In the cutting-edge frontiers of science, there is uncertainty that prevents claims gaining the status of knowledge.

Third, the attribution of knowledge is relative to a target system. We may generally ascribe the status of knowledge to models in scientific textbooks, because they correctly and accurately represent their idealized target systems. But such models may not have the status of knowledge when they are used for representing other target systems, such as real-world target systems, which may be messier or relevantly different from the textbook idealizations. In such a case, we would say that a model has internal validity, but not external validity. In other words, it's possible that a scientific representation would give us excellent knowledge of something, but poor knowledge of something else in which we are actually interested.

For example, in 1986, grazing lands in Cumbria, England, were contaminated with radioactive fallout. Radiologists confidently predicted that high radioactive cesium levels in the lands, its vegetation, and lambs who fed on it would fall soon, but their predictions repeatedly failed. It turned out that the model on which the scientists relied was developed from empirical observations of alkaline clay soils, in which cesium is chemically adsorbed and immobilized and so is unable to pass into vegetation, but in the acid peaty soil of Cumbria, cesium remained chemically mobile and taken up and returned to the soil by plant roots and animals that fed on the plants (Wynne 1992, 286–287; see also Section 5.2).

Another class of easy cases are genuine inter- or intradisciplinary scientific disagreements. In such cases, we may *not* legitimately attribute knowledge to the community. As the examples of the controversy between nineteenth-century chemists and the current controversy between generative linguists and psycholinguists (Section 3.3.3) illustrate, when the science isn't settled, the *Collective Justification* condition isn't met. I stress that the controversy must be *genuine* (Martini and Andreoletti 2021). Particularly, the mere existence of a stubborn dissent doesn't disqualify the majority view from constituting knowledge. I return to this point in Section 4.6.

How do hard cases look like? In hard cases, rightly or wrongly, the science is perceived as unsettled. The interference and entanglement of social factors, interests, and stakes make it harder for researchers to reach true claims and justify them against all possible objections and challenges.

Hard cases typically have the following characteristics. They involve inconvenient claims, the acceptance of which entails actions that some people prefer not to take, such as reducing carbon gas emissions, wearing medical masks, or getting them or their children vaccinated against coronavirus. They involve claims about which interested bodies, such as pharmaceutical companies or oil companies, have a lot to gain or to lose. They involve politically contested claims, especially theories that have become a battleground between political camps. They are cases, such as the theory of evolution by natural selection, in which scientific claims to knowledge are in tension with nonscientific epistemic authorities, such as institutionalized religion or state ideology. They are cases in which scientific predictions are for long-tern trends, such as a gradual temperature change over a century, rather than precisely defined observable events, such as a meteor shower. They are cases in which getting right answers and validating involve complex protocols, such as large clinical trials. They are cases in which scientific claims seem counterintuitive or in tension with people's personal experience, such as a person who believes that a homeopathic drug, which science claims is useless, has helped them, or that a medicine that science claims is safe, has led to a serious side effect in them.

Social media platforms have complicated the problem of knowledge attribution. Theories that were considered as more than settled, such as that the earth is round, are being challenged. Alternative epistemic authorities, such as influential bloggers and Twitter celebrities, gain many followers. Communities are formed around far-fetched theories, and their members, playfully or seriously, provide an endless stream of arguments to support them. These communities and arguments are easily found by search engines and appear side by side with information from established scientific authorities (Weinberger 2011). People share information from various sources, but the norms that govern sharing aren't stable across contexts and don't include inherent epistemic checks. When people read posts, they fill in information gaps drawing on their background beliefs, they understand the shared content differently, and attribute different levels of certainty to it (Record and Miller 2022).

Recall: the question here isn't what knowledge is (this question was answered in the previous section), but *how we can tell, in practice, who possesses knowledge or what constitutes knowledge in a noisy and epistemically polluted environment*. The next two subsections address this question, focusing on the question of when a consensus is knowledge based.

4.2 "Do Your Own Research"

It's *not* feasible for a person to evaluate whether an expert testimony or scientific consensus is knowledge by directly examining whether the testimony or consensual theory is supported by the available empirical evidence. If one could do that, the question of whether to trust an expert testimony or scientific consensus would not arise at all. Nonscientists typically lack the required competence to assess the evidence and cannot easily acquire it. Even a scientific insider typically lacks the time, resources and competence to evaluate all the available evidence and make an informed judgment. Scientific research is divided into many areas and subareas of expertise, and the amount of evidence produced is so vast and diverse that no one researcher or even a small group of researchers can evaluate it on their own. An individual who wants to reliably evaluate a scientific consensus cannot simply substitute the collective assessment of the scientific evidence by the scientific community with their own assessment.[31]

---

[31] On the prospects of individuals' "doing their own research" see Ballantyne et al. (2022), Buzzell and Rini (2023), and Levy (2022).

A similar proposal, which also leads to a dead end, is to evaluate whether the scientific claims were produced by employing the scientific method. There are three problems with this suggestion. *First*, historical, philosophical, and sociological studies of science militate against the view that there is such a thing as *the* scientific method. Scientists have been debating and using a variety of methods, of which some are more general and some are domain specific (Oreskes 2019, ch. 1; Knorr-Cetina 1999). While it's common in scientific textbooks to devote an introductory section to *the* scientific method, textbooks in different disciplines differ in their characterization of it and crudely patch together incompatible accounts as if they were all describing the same thing (Blachowicz 2009). There is no one description that unites all the methods used in all the sciences, or if there is, it's too broad and unspecific to be useful for our purpose. *Second*, methodology isn't independent of theory. Substantive expertise is required to know whether a methodology was correctly used. Thus, we return to the previous problem. *Third*, as we have seen in the previous sections, generating reliable scientific results requires not only that scientists adhere to methods, narrowly construed as technical procedures, but that they follow appropriate social-epistemic norms of inclusion and criticism. Therefore, evaluating scientific claims based on method alone overlooks relevant factors that are responsible for the reliability of scientific knowledge.

Perhaps one can evaluate a scientific consensus by examining whether the theory in question exhibits theoretical virtues? In her earlier work on consensus, historian of science Naomi Oreskes took this approach. Oreskes (2007, 79–92) identifies five general hallmarks of a good scientific theory: inductive support, predictive success, resistance to falsification attempts, consilience of evidence (convergence of evidence of different types), and explanatory success. Oreskes argues that because the theory of anthropogenic climate change bears these hallmarks, the scientific consensus over it is knowledge-based.

This approach, however, doesn't utilize the fact that scientific knowledge is social. It doesn't examine social indicators that may tell us whether the theory rises up to knowledge. Social evidence such as the number of people who support a view, their credentials, and their social background *is* relevant to assessing the credibility of a consensus. For Oreskes (2007), the existence of consensus, as a social fact, doesn't constitute a reason for (or against) trusting it. For suppose that the theory of anthropogenic climate change that currently enjoys a consensus were supported by a minority of the scientific community, and a rival theory, which doesn't bear Oreskes' five hallmarks of good science, were supported by the majority. By Oreskes' reasoning, the minority theory would still be more credible, because it would still be better evidentially supported. (Oreskes has since revised her criteria to include reference to social factors; see Section 4.4).

4.3 "Trust the Science"

An alternative approach that does take the social nature of scientific knowledge into account recommends that we "trust the science;" namely, trust a consensus when it's *scientific*. Those who subscribe to it aren't naive. They acknowledge that science is influenced by biases and interests, and that scientists are imperfect humans who produce imperfect results. Nevertheless, they hold that science is the most reliable way to the truth, thus laypeople have no better alternative than trusting a scientific consensus.

A key defender of this view is sociologist of science Harry Collins. Collins was once associated with the claim that the consensual outcome of scientific controversies is contingent on local social factors. But nowadays he argues that a scientific consensus, when such exists, should be trusted because the institution of science employs the cultural practices and values most apt for empirical inquiry, and the ethos of science makes scientists more virtuous, disinterested, and honest (Collins 2010; 2014, 126–132).

Philosopher Elizabeth Anderson (2011) advocates a similar position. Anderson distinguishes between trustworthy and untrustworthy scientific expert testimony. Anderson's criteria examine whether the testifiers possess recognized scientific credentials, whether the testified content underwent peer review, and so on. In principle, Anderson acknowledges the possibility that science may be untrustworthy, but her assessment criteria only examine whether the testimony is proper science, rather than fringe or crackpot. Her criteria overlook the possibility that *proper science* may be untrustworthy in some cases or on some matters.

As we know, however, even proper science isn't always right, and the knowledge it generates is not always apt for the tasks at hand. "If the history of science teaches anything, it's humility. There are numerous historical examples where expert opinion turned out to be wrong" (Oreskes 2007, 66). A scientific community or committee may be caught in groupthink (Solomon 2015, 96–100). A group of experts may mask internal disagreements and present a façade of consensus to keep their professional reputation intact and maintain public trust in science (Beatty 2006). A consensus can also emerge as a result of unrelated biases or influences in the scientific community that all happen to push in the same direction (Miller 2013, 1306–1308; Solomon 2001, 121–135).

Because a scientific consensus may be wrong, partly wrong, or otherwise epistemically flawed, indiscernible deference to a scientific consensus is typically incompatible with fulfilling one's epistemic responsibilities. The question of whether a scientific consensus is knowledge based usually arises when a possible course of action is considered. Blind compliance with science is dangerous. Those who wish to rely on a scientific consensus ought to have sufficiently good reasons to think that the consensus view isn't wrong or flawed. This requirement is in line with the *Individual Responsibility* condition for knowledge (Section 3.3.2). Recall that this condition doesn't require that an individual assess on their own all the available evidence in the scientific community in order to have knowledge, but that they be satisfied that the evidence that supports the putative knowledge exists.

For instance, in the beginning of the COVID-19 crisis in 2020, the consensus among the experts who advised the British government was that restrictive measures such as lockdowns would be ineffective and impractical in the UK. The British government adopted this view and did not initially take such measures. A report by two House of Commons committees, overseeing health and social care, and science and technology criticizes the government for that:

> We accept that it is difficult to challenge a widely held scientific consensus. But accountability in a democracy depends on elected decision-makers taking advice, but examining, questioning and challenging it before making their own decisions (House of Commons 2021, §102).

It further notes:

> The fact that the UK approach reflected a consensus between official scientific advisers and the Government indicates a degree of groupthink that was present at the time which meant we were not as open to approaches being taken elsewhere – such as earlier lockdowns, border controls and effective test and trace – as we should have been. (House of Commons Health and Social Care, and Science and Technology Committees 2021, §152)

While the report authors have the unfair benefit of hindsight, their point is still valid: officials ought to take an inquisitive stance toward a scientific consensus and the reasoning underpinning it before they accept and act on it. The next subsection reviews ways in which this can be done.

4.4 Assess Whether a Scientific Consensus Is Knowledge Based

What might taking an inquisitive stance toward a scientific consensus involve?[32] I have previously developed an account of knowledge-based consensus that answers this question (Miller 2013; 2016). I identify four types of consensuses, of which three aren't knowledge-based and one is. The first is a noncognitive consensus, which aims at promoting nonepistemic aims. The second is a "vertically lucky" consensus. This is an agreement that happens to be correct, but could have easily been wrong. The third is an epistemically unfortunate consensus, in

---

[32] See also Miller (2019) and Dellsén (Forthcoming).

which parties to the consensus have the bad luck of being systematically or deliberately misled or biased. When a consensus belongs to none of these three types, it's likely to be knowledge-based.

I have argued that inasmuch as we can eliminate nonepistemic factors, veritic epistemic luck, and epistemic misfortune as the best explanations of a consensus, knowledge remains its best explanation. I have identified three conditions for knowledge being the best explanation of a consensus: (1) *social calibration*: researchers give the same meaning to the same terms and share the same fundamental background assumptions; (2) *apparent consilience of evidence*: the consensus seems to be built on an array of evidence that is drawn from a variety of techniques and methods; (3) *social diversity*: the consensus is shared by men and women, researchers from the private and public sectors, liberals and conservatives, and so on. These three conditions can be summarized as follows:

> h merits our confidence insofar as a socially diverse group of researchers, working within similar frameworks of fundamental assumptions, accepts h on the basis of different kinds of evidence that are apparently consilient (Rehg and Staley 2017, 132).

Oreskes has recently also developed an account of knowledge-based consensus. Oreskes (2019) draws on Longino's account of scientific knowledge (see Section 3.2) to argue that we can be confident in accepting a consensus view in a scientific community if the community is "sufficiently diverse, self-critical, and open to alternatives, particularly in the early stages of investigations when it's important not to close off avenues prematurely" (143). Specifically, Oreskes formulates a list of questions that should be answered in the positive:

- Do the individuals in the community bring to bear different perspectives? Do they represent a range of perspectives in terms of ideas, theoretical commitments, methodological preferences, and personal values?
- Have different methods been applied and diverse lines of evidence considered?
- Has there been ample opportunity for dissenting views to be heard, considered, and weighed?
- Is the community open to new information and able to be self-critical?
- Is the community demographically diverse: in terms of age, gender, race, ethnicity, sexuality, country of origin, and the like? (Oreskes 2019, 143–144)

Oreskes and I both stress social diversity in the scientific community as a condition for knowledge-based consensus (see Section 4.5). Our respective accounts, however, also differ. While Oreskes, following Longino, focuses on evaluating the social *processes* of knowledge generation and the norms that govern them, my account focuses on evaluating their *outcomes*.

I have previously argued that process-based accounts of knowledge-based consensus are inadequate because good processes don't guarantee good outcomes; hence it doesn't suffice to evaluate the processes that lead to the formation of a consensus without evaluating its outcomes (Miller 2019; see Sections 3.3.3 and 3.6).

I still think that this argument is basically correct, but I now see process-based and outcome-based approaches for assessing a consensus as complementary rather than competing. First, they focus on two different questions, which are both important. A process-based approach tells us how to reach a knowledge-based consensus, while an outcome-based approach tells us how to evaluate whether a consensus that has already been reached is knowledge-based. Second, when assessing a consensus in practice, the empirical facts on which a process-based and an outcome-based assessment of consensus would focus overlap. For example, if one wants to examine whether the evidence that supports a consensus view is consilient, one should look at how this evidence was generated, and whether alternative theories that can explain it were adequately considered.[33]

Having said that, it's important to highlight a controversial assumption that Oreskes' process-based account of knowledge-based consensus makes. The assumption is that the scientific community *aims* at achieving a consensus, specifically a *hard-won* consensus, namely, a consensus "that emerges only after vigorous debate and a thorough examination of the range of alternative explanations" (Ranalli 2012, 187).[34] In a hard-won consensus, even the maverick, skeptical members of the scientific community concede that the consensual view is well supported by the empirical evidence, which, in turn, gives us reasons to trust it.

It's debatable, however, whether science generally aims at consensus. Reviewing several examples from the history of science, Miriam Solomon concludes that the causes of consensus and dissent are the same: various contingent factors, both empirical and nonempirical, impel each individual

---

[33] Rehg and Staley (2008) make a similar point. They assess a consensus by the cogency of the consensus argument. One way to characterize cogency is as an objective relation between premises and a conclusion. Under this characterization, social procedures that lead to the acceptance of the argument and profiles of the people involved in the process are potential second-order evidence that the argument is cogent, as they may indicate that the argument attends to the total available evidence, and that the relations of support between the premises and the conclusion were critically examined. Alternatively, such social facts may be thought of as constitutive of cogency, namely, as conditions that must be met for a consensus argument to be cogent. Either way, in practice, assessing the cogency of a consensus argument requires examining the same social facts.

[34] The notion of "hard-won consensus" is attributed to Oreskes by Stephen Macedo in the introduction to her 2019 book (p. 6). This notion was originally developed and applied to climate science by Ranalli (2012).

member of the scientific community to subscribe to one theory or another. No scientist is exposed to the whole picture. Thus, every scientist forms their own judgment based on the partial picture they see. Their training and experience cause each scientist to be more persuaded by certain kinds of evidence and arguments than by others. Each individual is also subject to a different influence of social factors such as ideology and religion. In aggregate, these factors sometimes produce dissent and sometimes produce consensus. For Solomon, because both consensus and dissent emerge due to the aggregation of accidental factors, dissent isn't a temporary glitch to be overcome, and consensus isn't the end of inquiry (Solomon 2001, 117).

If Solomon is right, a consensus, even a hard-won one, cannot be simply explained by the telos of science, namely as a result of scientists' following the norms of scientific inquiry, since these norms don't necessarily lead to a consensus. Besides reviewing the norms and practices that generated the scientific consensus, we should still ask: what is the best explanation of the consensus? Is it bias, aggregation of biases, silencing of dissenting views, pressure to present a unified front – or is it knowledge?

4.5 The Epistemic Importance of Social Diversity

Both accounts of knowledge-based consensus I reviewed stress the importance of social diversity. How does diversity help achieve knowledge?[35] Diversity has many epistemic benefits. Diversity may generate new research questions, identify limitations within existing models, propose new models, propose alternative hypotheses and interpretations of data, open up new lines of evidence, reveal "loaded" language in descriptions of phenomena, and more adequately identify and weigh potential risks (Intemann 2009). Diversity raises the chances that marginalized or overlooked positions associated with members of relevant weak groups get proper consideration. If one of these otherwise neglected views is true, diversity increases the chances the epistemic community will consider and accept it. By doing so, it increases the chances that the *Veracity* condition is met.

Diversity also helps satisfy *Collective Justification* and *Objective Justification*. As mentioned in the discussion of these conditions, a central problem in theory choice is a failure to conceive of alternative explanations. If for a putatively successful theory T, it very easily could have been the case that had we thought of an alternative T^*, we would not have accepted T, then if T is true, we are lucky to have accepted it. Social diversity increases the number of alternatives we consider; hence, we can be more confident that if T is true, it's not merely luckily true (Miller 2013, 1313).

---

[35] This section draws on Miller and Pinto (2022).

Two lines of objection may be raised against these claims. According to the first, the epistemic drawbacks of social diversity outweigh its epistemic benefits. Second, social diversity is merely a means of achieving other epistemic aims, which can be reached more effectively without diversity.

Start with the first line of objection. One may worry that including outsiders, who have fundamentally different background assumptions and epistemic commitments and are unconstrained in their evidence assessment by accepted standards, in a scientific community might inhibit the formation of warranted agreement. Considering too many alternative theories may distract researchers and reduce the chances that a true theory will eventually prevail (Stegenga 2016, 46). Social diversity has additional epistemic drawbacks. Different epistemic communities may mean different things by the same terms, which may lead to mutual misunderstanding. They may have different certainty norms. For example, members of some communities may make reports even when they are uncertain, while others may make reports only when they are certain. This makes members of different communities misevaluate each other's credibilities or miss out on information (Gerken 2022, 248–257).

Indeed, social diversity, particularly including relevant uncredited experts and stakeholders in a knowledge generation discourse, isn't an epistemic panacea. It has epistemic pros and cons. Empirical evidence indicates that social diversity generally improves collective epistemic performance, but measures must be implemented to mitigate obstacles that come with diversity, such as intergroup friction (Page 2017, ch. 5). The difficulties with diversity, however, aren't unique to interactions between scientists and outsiders, but exist in interdisciplinary collaborations. Scientists have developed effective ways to overcome these problems (Miller and Freiman 2020), which can also be used in interactions with outsiders.

One may also worry that too diverse a group may hold conflicting, hence incoherent, background assumptions and reasons to support a consensus view, thus failing to achieve *Collective Justification*. But as Haxin Dang (2019) argues:

> When members of an epistemic group disagree over the reasons for their claim, this disagreement is not immediately bad. If this disagreement is a result of negotiating different background theories or resolving multiple sources of evidence, then the epistemic group ought to be nonetheless justified in its conclusion. Diversity in these cases is epistemically valuable and should be maintained, not eliminated. (1035)

Dang argues that agreement should be over the research question (similarly to my *social calibration* condition, Section 4.4) and over the result or conclusion. Intermediate disagreements should themselves be justified, namely, reason supported.

The debate about diversity is reminiscent of the debate between Thomas Kuhn and Karl Popper about the necessary conditions for the growth of knowledge. Kuhn (1970) emphasized the importance of a conceptual framework (paradigm) to define research problems and standards of accepted solutions to them, focus researchers' attention, and shield them from distractions in the form of attacks on, and doubts about their fundamental theoretical commitments. By contrast, Popper (1970) emphasized the necessity of constant criticism of, and challenges to researchers' fundamental theoretical commitment, lest science become a dogmatic enterprise of perpetuating and reaffirming received views regardless of their empirical validity. Assuming that social diversity is needed for bringing perspectives that challenge accepted dogmas, then for Popper lack of diversity is a serious epistemic problem.

Both a guiding conceptual framework, as Kuhn argued, and challenges to its core assumptions, as Popper argued, are required for the growth of knowledge. Their exact positive and negative contributions would depend on the case at hand. A question remains, though, whether social diversity is *necessary* for achieving plurality of perspectives and the other epistemic benefits associated with diversity. According to the second line of objection, which states that social diversity is merely a means of achieving other epistemic aims, it's not. Social diversity (aka "identity diversity") is epistemically beneficial because it's correlated with cognitive diversity, which is diversity in problem-solving methods and strategies. In the context of justification, additionally to generating alternative hypotheses, cognitive diversity leads to the generation of evidence of multiple types. When such multimodal evidence points in the same direction, it provides greater confirmation of a hypothesis than evidence of a single type. This is the idea of robustness (Stegenga 2009). Social diversity is thus merely a means of achieving cognitive diversity, which in turn is a means of achieving theoretical pluralism and evidential robustness. Cognitive diversity, however, can be achieved by other means, such as teaching a socially homogenous group to use a variety of research methods – or so the second objection goes.

The reply to this objection is twofold. First, achieving cognitive diversity by means other than identity diversity is often unfeasible. Identity properties are correlated with different skills, bodies of knowledge, and mental models for analogical reasoning, and identity homogeneity is correlated with cognitive homogeneity. Homogenous groups may be unaware of their knowledge gaps. Even in narrowly construed technical matters, social diversity is correlated with different training methods. People from different social backgrounds go to different colleges. Some colleges emphasize abstract mathematical skills, while others take a hands-on approach. Colleges don't follow exactly the same curriculum. Closing such wide, potentially unknown gaps of knowledge and skills in a homogenous group is often impractical (Page 2017, 149–161).

Second, the positive contribution of social diversity is distinct from that of robust evidence. Even if a scientific community accepts a theory that seemingly enjoys strong evidential support, social diversity in the community is an independent epistemic desideratum. Suppose there is a scientific consensus that passive smoking doesn't raise the chances of lung cancer. Suppose this conclusion is supported by seemingly robust studies of different types, such as epidemiological, in vivo, and in vitro studies. If all these studies have been supported by tobacco companies, there is still a good chance their conclusion is false. The upshot is that for inferring truth from social agreement, in addition to being evidentially supported, we would like it to be socially diverse in relevant ways, in this case shared by publicly funded researchers, smokers and non-smokers, and so on. It may be objected that the consilience of evidence here is merely apparent, and not genuine. Hence, this example doesn't show that social diversity trumps evidential robustness. Be that as it may, social diversity, or awareness of lack thereof, is still necessary for discovering that the evidence isn't as robust as it seems (Miller 2013, 1312).

Menachem Fisch (2017) explains why identity diversity has an indispensable role in scientific knowledge generation. Fisch takes seriously the possibility of incommensurability between the normative epistemic standards of two competing conceptual frameworks.[36] He argues that there are limits to an insider's ability to criticize or challenge a framework from within that framework, because the insider is limited in his ability to criticize the very standards on which his criticism is based. Only an outsider, who adheres to a different conceptual framework, and isn't committed to the insider's standards, can thoroughly criticize the insider's framework. The outsider's perspective is indispensable for the successful generation of knowledge.

But because the outsider argues from within a different framework, so Fisch argues, she cannot rationally convince the insider. The outsider can only destabilize the insider's beliefs and make him reflect on his commitments. But in order to do that, the insider must *trust* the outsider. Namely, he must take her to be offering a system of commitments that potentially constitutes a genuine alternative to his own. The insider must treat the outsider *as an epistemic equal*,[37] whose

---

[36] See footnote 19.
[37] See Miller and Pinto (2022) for a full account of the meaning, importance, and implementation of epistemic equality.

contribution is potentially as valuable as his own. Hence, incorporating outsiders and treating them with equality is epistemically indispensable.

Additionally, marginalized outsiders may have knowledge that privileged insiders lack. The servants know the house better than the landlords because of their hands-on experience maintaining it and their social transparency, which allows them to be exposed to secrets (Wylie 2003). Similarly, public transport users know better than public-health experts who use private transportation which coronavirus restrictions on public transport are realistic.

4.6 Appropriate and Inappropriate Scientific Dissent

So far, I argued that when some conditions obtain (social diversity being an important one), we may legitimately attribute knowledge to an epistemic community that has reached a consensus, but we may not legitimately attribute knowledge to an epistemic community when a genuine scientific controversy on the relevant matter exists. One difficulty with this argument is that a complete scientific consensus almost never exists. Unanimity in science is rare. Unanimity even constitutes a pro tanto reason *against* attributing knowledge to the consensus. This is because in pluralistic settings, even when the evidence is compelling, we should still expect some people not to be persuaded by it. When this isn't the case, a suspicion arises that silencing of views occurred (Dellsén 2021).

So, the question to ask is when the existence of dissent indicates the existence of a genuine controversy in a scientific community, and when it does not. Another way to put this question: when is a dissent normatively appropriate? When does a dissent play a positive role in the growth of scientific knowledge? When a dissent is normatively appropriate, it may help either fortify the consensus view against objections or eventually overturn it. It indicates that a matter has not been adequately settled. In contrast, when a dissent is normatively inappropriate, it "fails to yield any of the epistemic benefits that make even false dissent valuable" (de Melo-Martín and Intemann 2018, 16).

A normatively inappropriate dissent is epistemically detrimental. It hinders the growth of scientific knowledge by preventing warranted closure of scientific controversies, and by leading community research and argumentation efforts astray in unfruitful directions. A normatively inappropriate dissent also confuses the public and decision makers about policy-relevant science, such as the theory of anthropogenic climate change. When a dissent is normatively inappropriate, it can safely *not* be taken into consideration when we assess the consensus view for knowledge.

Four accounts of normatively inappropriate dissent may be identified in the literature. One account is provided within Longino's Critical Contextual Empiricism (section 3.2), according to which a dissent is normatively inappropriate when the dissenters don't follow the norm of uptake of criticism; namely, when they don't respond in a relevant and appropriate way to criticism of their view and continue to hold the dissent view regardless.

As de Melo-Martín and Intemann (2018, ch. 4) and Emanuela Fernández Pinto (2014), argue, however, Longino's account of normatively inappropriate dissent faces difficulties dealing with manufactured uncertainty. Dissenters usually do present arguments for their view, and it's hard to tell whether their responses are adequate, or irrelevant and repetitive. Namely, in practice, it's hard, especially for outsiders, to judge whether dissenters fail to follow the norm of uptake of criticism and are just being stubborn, or whether their concerns are genuine. Therefore, it's hard to determine when a community may stop engaging with them and move on.

Defending Longino against this anticipated charge, Borgerson (2011, 445) argues that if we distinguish the level of certainty required for action from that required for knowledge, interested parties will be less motivated to manufacture uncertainty. In response, I have argued elsewhere that Critical Contextual Empiricism should still be able to determine when closure in an epistemic community is warranted despite incessant criticism (Miller 2015, 118–119).

According to a second suggestion, defended by Oreskes (2019, 65–68), dissent is normatively inappropriate if the dissenting scientists act in bad faith and have a conflict of interest in that the acceptance of the consensus view may financially hurt them or their industry sponsors. As de Melo-Martín and Intemann (2018, ch. 3) argue, however, we don't have access to scientists' mental states, therefore it's hard to assess scientists for their motivations. With respect to conflict of interests, Oreskes admits that industry-funded science can still be valuable "so long as the norms of critical interrogation are operating and conflicts of interest are forthrightly disclosed and where necessary addressed" (2019, 66–67). But if the criterion for normatively inappropriate dissent is ultimately whether the dissenters follow norms of critical scrutiny or not, then Oreskes' account boils down to Longino's account, which, as argued earlier, is unsatisfactory. To be clear: I don't deny that acting in bad faith or having a conflict of interests constitutes a reason against trusting scientists' claims. The question here is how to integrate this truism into a philosophically satisfying and practically useful account of normatively inappropriate dissent.

Biddle and Leuschner (2015) provide an inductive-risk account for distinguishing epistemically beneficial dissent from normatively inappropriate dissent. Drawing on Wilholt (2009), who characterizes conventional scientific epistemic standards, for example, using a critical p value of 5 percent, as reflecting conventional trade-offs between inductive risks, Biddle and Leuschner formulate four conditions jointly *sufficient* for a dissent over a hypothesis H to be considered normatively inappropriate dissent:

Inductive-Risk   The nonepistemic consequences of wrongly rejecting H are likely to be severe.
Standards   The dissenting research that constitutes the objection violates established conventional standards.
Producer-Risks   The dissenting research involves intolerance for industry-producer risks at the expense of public risks.
Different-Parties   Producer risks and public risks fall largely upon different parties.

A paradigmatic case of normatively inappropriate dissent that meets Biddle and Leuschner's criteria is the long-lasting dissent from the mid 1950s to the mid 2000s denying a causal connection between smoking and lung cancer, and the harms of passive smoking. This dissent was based on contrarian research and advocacy heavily funded by tobacco companies (Oreskes and Conway 2010). The consequences of this prolonged dissent were illness and death to smokers and people in their close vicinity (conforming to *Inductive-Risk*). It relied on dubious scientific standards (conforming to *Standards*). It brought about large profits to tobacco companies at the expense of public health (conforming to *Producer-Risks*). And the public, rather than tobacco companies, carried most of the burden of these expenses (conforming to *Different-Parties*).

de Melo-Martín and Intemann (2018) make a twofold criticism of Biddle and Leuschner's inductive-risk account of normatively inappropriate dissent. First, they argue that conditions *Producer-Risks* and *Different-Parties* that address the distribution of inductive risks are inadequate because some clear cases of normatively inappropriate dissent fail to meet them. Both the dissent that denies that the HIV virus causes AIDS and the dissent that questions the safety of childhood vaccines don't involve intolerance for producer risks at the expense of public risks. In these cases, the dissent goes against the financial interests of drug companies that produce vaccines and AIDS treatments (66). Strictly speaking, this example doesn't refute Biddle and Leuschner's account, because it specifies sufficient, rather than necessary conditions for normatively inappropriate dissent. But its failure to capture a clear case of normatively inappropriate dissent significantly hinders its usefulness.

Indeed, whether a dissent is intolerant to producer risks depends on whether the producers produce a problem or a solution. While conditions *Producer-Risks* and *Different-Parties* arguably fail, the rationale behind them is still sound. They are meant to prevent an unjust, unfair, or illegitimate weighing or distribution of inductive risks that would stem from accepting the dissent view.

To overcome these difficulties, I have suggested (Miller 2021b) the following three conditions, which, if a dissent *fails to meet*, would make it a normatively inappropriate dissent:

Political-Legitimacy	The dissent view has been generated, and adopted as a basis for public action – it if has been adopted – by a politically legitimate knowledge-producing and stabilizing process.
Fairness	The inductive risks that would stem from accepting or adopting the dissent, and the respective harms and benefits that they entail, would be fairly and justly distributed among relevant members of, or groups in society.
Workable-Evidential-Thresholds	The dissent adopts evidential thresholds within a range which allows researchers to make meaningful knowledge claims in a timely manner,[38] and which isn't knowingly likely to rule out a priori the attainment of certain empirical results.

These first two conditions ensure that both the *process* and *outcome* of trading off inductive risks against each other are legitimate (see Section 2.4). With respect to *Political-Legitimacy*, the process in question is not only the internal scientific process of research and validation, but also the external process of considering certain claims as a basis for public action. As I argue in Section 5.5, the stabilization and acceptance or rejection of knowledge claims in policy contexts depends not only on their epistemic validity, narrowly construed, but also on the perceived political legitimacy of the process of their consideration. The *Political-Legitimacy* condition turns this insight into a normative requirement. It would impose restrictions on the process, such as a prohibition on massive lobbying, and adequate public and stakeholder representation.

The *Workable-Evidential-Thresholds* condition ensures that conventional evidential thresholds are within a range that allows scientists to make meaningful knowledge claims, for example, to distinguish signal from noise or genuine events from chance events. Additionally, the evidential standards shouldn't be set in a way that prevents or makes it very unlikely or to reach a certain outcome (see the corn and ladybugs example in Section 5.2).

4.7 Conclusion

This section addressed legitimate knowledge attribution in real-life cases (see Figure 1 for a section summary). When there is a near knowledge-based consensus

---

[38] The words "in a timely manner" were added following criticism by Hicks (2021, 9).

in a scientific community, possibly excluding only a normatively inappropriate dissent, we may attribute knowledge to the scientific community. By contrast, when there is a normatively appropriate dissent or when the issue isn't internally settled, we may not attribute knowledge to the community. In such cases, it's better for involved parties, particularly official bodies, to acknowledge uncertainties rather than to pretend to conclusively know. For example, early in the coronavirus pandemic, the WHO claimed that it was an established fact that COVID-9 was *not* airborne, only to quietly retract that claim two years after (Lewis 2022). Even when we may legitimately attribute knowledge, we should still ask whether this knowledge is full and adequate for our purposes, which is one of the issues the next section addresses.

5 The Interaction of Epistemic and Social Elements in the Coproduction of Knowledge

The conception of knowledge developed in Section 3 asked how epistemic and social elements fit together in the conditions for scientific knowledge. Section 4 asked what identifiable indicators there are that these conditions for scientific knowledge are met in a given case. This section addresses a complementary

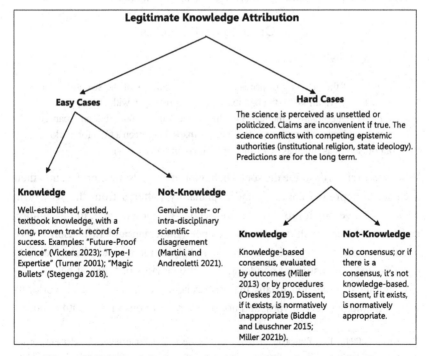

Figure 1 Section 4 Summary: Legitimate Knowledge Attribution

question: how do epistemic and social elements interact to produce scientific knowledge?

The idea that knowledge is produced by social and nonsocial elements is captured by Sheila Jasanoff's idiom of coproduction. According to Jasanoff (2004a),

> Knowledge and its material embodiments are at once products of social work and constitutive of forms of social life; society cannot function without knowledge any more than knowledge can exist without appropriate social supports. Scientific knowledge, in particular, is not a transcendent mirror of reality. It both embeds and is embedded in social practices, identities, norms, conventions, discourses, instruments and institutions – in short, in all the building blocks of what we term the *social* (3; emphasis in the original).

This section analytically reviews the mechanisms coproduction; namely, the modes of interplay between epistemic elements (representations, data, evidence, etc.) on the one hand, and social elements (values, interests, ideologies, etc.) on the other hand.[39]

In what follows, the terms "values" and "interests" are interchangeable, as they are both understood as someone's discriminators between more or less preferred states of material, biological, social, political, and cosmic orders. A worldview that is based on such values, for example, a worldview that disapproves of abortions or supports same-sex marriage, I will call "ideology."

5.1 Ontological Projection

Ian Hacking (1999) writes:

> accounts of the behavior of primates reflect the societies of the scientists who study them. We all know the bad jokes about British apes with stiff upper lips, ruthlessly enterprising American apes, hierarchical and communitarian Japanese apes, promiscuous French apes. Primates, perhaps, have been a field for working out ourselves as much as describing animal communities. (64)

When researchers describe the social behavior of animals, they project into their accounts interpretive concepts and explanatory patterns from the society in which they live. Such projection is bidirectional. For example, in our culture, and even in areas such as business science (Ludeman and Erlandson 2004), people borrow ethological terms such as "alpha male" and "beta male" to describe, excuse, justify, or condemn certain behaviors or types of people.

Ontological projection means that researchers project their social world, its values, ideologies, stereotypes, and normative perceptions into scientific models

---

[39] Jasanoff (2004b) lists four axes of interaction between social and nonsocial elements in the coproduction of scientific knowledge: identities, institutions, discourse, and representations (39–43). The following analysis parses the space of interactions slightly differently.

and the language they use to describe them. Ontological projection involves both metaphors and *analogical reasoning*,[40] which can be potent scientific tools of inquiry. Ontological projection works both ways: our culture also projects scientific models into the social world, extending their original domains.

Ontological projection isn't limited to ape tribes, bee colonies, and packs of wolves. For instance, various social systems have been used to describe and model the brain. Economists have modeled the brain as a "hierarchical organization" (Brocas and Carrillo 2008); an academic accountant describes the brain as "the original accounting institution" (Dickhaut 2009); a central theory of cognitive development likens the brain to science (Gopnik and Meltzoff 1998); network scientists view the mind as a network (Hills and Kenett 2022), and with the rise of blockchain technology, it has been suggested that the brain is a "decentralized autonomous corporation" (Swan 2015).[41]

Ontological projection exists in biology too. American science textbooks commonly describe the human body and its immune system as a nation state at war over its external borders (Martin 1990). A well-studied example in feminist science studies is the "Prince Charming/Sleeping Beauty" model of the egg and sperm. Textbooks and research papers have commonly described fertilization in gendered terms, portraying the sperm as an active hero who awakens and rescues a passive and submissive "damsel in distress" egg. Namely, biologists have projected stereotypical perceptions about gender roles in society into a biological model. This projection has made scientists overlook the role of the egg in the process of fertilization, which has resulted in a knowledge gap. It has also led to errors: it was widely claimed that the mechanical force generated by its tail causes the sperm to penetrate the egg. Evidence suggests, however, that this force is insufficient to penetrate the egg, that the egg chemically immobilizes sperms that come in contact with it, and that external egg organelles bring the sperm into the egg (Beldecos et al. 1988; Martin 1991).

Another example comes from mathematics. Roy Wagner (2009) describes how gendered language has been used in the formulation and proof of the "stable marriage" theorem. Although the theorem (like any mathematical problem) can be stated in abstract mathematical terms, it's commonly presented in gendered terms, and its algorithmic proof is constructed as a narrative about men who actively court passive women. The original context in which the algorithm was developed was matching medical interns to hospitals. Wagner argues that the use of gendered language has blinded mathematicians to some of the theorems' mathematical implications.

---

[40] For an argument to the effect that metaphors exceed analogical reasoning see Hesse (1988).
[41] I thank Axel Gelfert and Gen Eickers for these examples, which are all from mainstream scientific journals.

As mentioned, ontological projection can go in the opposite direction, from nature or mathematics to society. The field of "sociophysics" or "social physics" uses models from physics, such as statistical mechanics and fluid dynamics models, to represent and study social phenomena such as voting (Galam 2012; Pentland 2014).

As Fox-Keller (2015)[42] argues, ontological projection can also occur between scientific fields, such as in taking genes to be the "atoms" of life (physics/chemistry to biology) or the genome to be the "program" of the organism (computer science to biology). The 2000s' shift from referring to non-coding-DNA as "junk DNA," which implied that it plays no role in the organism and that all the functions of DNA can be reduced to coding proteins, to referring to it as "the dark matter of the genome," has transformed it into a riddle for biologists just like dark matter is a riddle for physicists. These metaphors are powerful in that they shape researchers' theoretical expectations, open up, and close down research avenues (see Sismondo 1996, 143).

Brendon Larson (2011) describes a form of ontological projection, which he calls "feedback metaphors." Feedback metaphors are "scientific metaphors that harbor social values and circulate back into society to bolster those very values" (22). Feedback metaphors constitute central theoretical terms in their respective scientific domains, but also carry with them normative outlooks that favor certain modes of thinking, types of solutions, and courses of action. For example, the biological term "invasive species" invites us to think about a species in militaristic terms as an enemy that needs to be fought off (Larson 2011, ch. 6).[43]

In ontological projection, then, there is a reciprocal process of normative reinforcement between scientific models and the values and ideologies that are projected onto them. While apt scientific metaphors facilitate understanding of and allow justified inference about their targets (Levy 2020), inapt, unreflective ontological projection may unjustifiably boost the perceived validity of knowledge claims. This is because the perceived validity of knowledge claims depends on our stance toward the values we project. In the other direction, models unjustifiably lend scientific credibility to the social values and ideologies projected onto them. The projection of commonly held and infrequently questioned social stereotypes onto the egg-and-sperm model has strengthened its perceived validity, and vice versa: the egg-and-sperm model reinforces the gender stereotypes projected onto it by suggesting that they are natural and biological in origin.

Might the projection of "good" values, for example, egalitarian gender values, onto scientific models promote the generation of better knowledge? The answer is a qualified "yes." The qualification is that social values have no evidential power.

---

[42] I thank Lee Nelson for this reference.   [43] I thank Kevin Elliot for this reference.

Therefore, whether they are "good" or "bad" values neither strengthens nor weakens the validity of knowledge derived from, or constituted by the models onto which they are projected. Yet thoughtful, reflexive, and responsible ontological projection is still epistemically beneficial. In the context of discovery, projecting certain values rather than others may unblind researchers to certain research directions and lead them to formulate fruitful hypotheses they would not have formulated otherwise. The more hypotheses are explored, the less likely it is that the hypothesis that eventually prevails will prevail due to bias (Okruhlik 1994).[44] When a certain metaphor is associated with certain policies, such as with waging war on invasive species, an alternative metaphor might point at alternative, perhaps preferable, solutions. If we construe the context of justification widely to include considerations of political legitimacy and inductive risks (as the conception scientific knowledge in Section 3 does), then broadening the space of action possibilities envisioned by a model contributes to its epistemic standing (see Figure 2 for the section summary).

5.2 Social Values Affect Representation Decisions

Social values and interests affect researchers' representation decisions. Social values and interest affect two kinds of representation decisions: (1) *what* to represent – what entities to include in the model; (2) *how* to represent (Harvard

---

Ontological Projection

A form of analogical reasoning in which structures are projected:

- from society to scientific models of nature (or mathematics);
- from scientific models of nature to society.

When projected, the structures carry with them values and normative perceptions from their original domain.

Projected social values ⟷ **Scientific models**

mutually reinforce the perceived validity of

Depending on their content, projected values may open up or close down fruitful research avenues or alternative hypotheses.

By themselves, projected values have no evidential force. They neither strengthen nor weaken the epistemic standing of the models to which they are projected.

Figure 2 Ontological projection

---

[44] See also Section 5.2.

and Winsberg 2022; Harvard et al. 2021). Concrete claims or predictions that are derived from models have truth-values, but according to a common view in philosophy of science, models themselves are neither true or false, but rather adequate or inadequate for purpose.[45] This means that the normative evaluation of the effect of social values on representation decisions not only addresses the question of whether these decisions lead to true claims or predictions, but also whether they serve the epistemic and social aims of the model. Put differently, in representational decisions, we don't only care whether they lead to knowledge rather than error, but also whether they lead to useful, rather than useless, knowledge (Parker 2020).

Values affect decisions about *what to represent in a model*. An example of a representation decision is the choice of a comparator in a drug clinical trial. The researchers may test the drug against a placebo, against a substandard comparator, or against the standard treatment. Social interests play a role in the choice. Pharmaceutical companies want to get the drug approved while making it seem as efficacious as possible. Therefore, they have an interest to choose the most inferior comparator they can get away with. By contrast, prospective drug users have an interest in the new drug being compared to the standard treatment already available in the market in order to know whether the new drug is any better. Other decisions that are subject to the influence of interests in clinical trials are participants' eligibility criteria and selection end point. Pharmaceutical companies have an interest that the trial population would be selected such that the efficacy of the drug would be clearly demonstrated. This typically means that it would consist of patients who suffer from no other condition and take no other drugs, which can interfere with the one tested. By contrast, prospective drug users have an interest in the trial population being as similar to them as possible (Harvard and Winsberg 2022, 17–20).

Harvard and Winsberg (2022, 17–20) argue that on their own, inadequate decisions about what to represent don't directly lead to error, since they merely are about what to include in the model. Hence, the harms they can cause cannot be conceptualized in terms of inductive risk management (Section 2.4). Rather, the hazards that are associated with inadequate representation decisions are getting *incomplete*, *irrelevant*, or *distracting* results. For example, analyzing the AstraZeneca (AZ) COVID-19 vaccine rollout model, Harvard et al. (2021) note that the AZ researchers' decision to represent subjects by age, sex, and "front-line worker" status, but not by their race, income, postal code, occupation, or household size made the AZ model

---

[45] See Harvard and Winsberg (2021, 2) for references; see Perini (2012) for a counter view.

not adequate for exploring research questions like i) the societal-level effects of structuring the AZ vaccine roll-out by race, income, postal code, occupation, or household size; ii) the individual-level effects of race, income, postal code, occupation, or household size on the personal harm-benefit ratio of immediate vaccination with AZ versus delayed vaccination. (2)

Nevertheless, an inadequate representation decision may lead to error if it violates scientific standards that require that the represented entities have certain properties that would allow inferring or extrapolating from the model to other entities. Such errors *can* be conceptualized in terms of inductive risk management. For example, industry-funded studies that were supposed to test whether exposure to bisphenol A causes health problems used a strain of rat that is particularly insensitive to any estrogen. As expected, they found that bisphenol A was not associated with health problems in these rats (Wilholt 2009, 93). But the prevailing scientific standard requires that the choice of experimental organism allow us to extrapolate from it to other species, particularly humans. Since this standard was violated, an extrapolation from the model may lead to error. Another example: in 2001, an industry-employed researcher performed an experiment on corn that was genetically modified to contain a toxin for combating rootworms. After eight days of experiment, almost 100 percent of ladybugs that ate the corn died. In a subsequent experiment, the researcher conveniently stopped recording data after seven days, namely, made a decision about how many days to represent. The researcher then inferred that the corn was generally safe for ladybugs (Biddle 2014, 18). This illegitimate inference that leads to error is a result of an inadequate decision about what to represent.

A second family of representation decisions is about how to represent. One decision about how to represent concerns *the sources(s) of data*. For example, should the AZ COVID-19 vaccine model represent the risk of mortality from vaccine-induced immune thrombotic thrombocytopenia as 1 in 153,000 with a 21 percent chance of death, as estimated by the European Medicines Agency, or as 1 in 100,000 with a 40 percent chance of death as estimated by Canada's National Advisory Committee on Immunization? Should it average them, or use both? (Harvard et al. 2021, 3). Making the right decision must involve making value judgments about the different inductive risks associated with each decision.

Decisions about how to represent also concern how to *(1) individuate variables, (2) parse a variables space, and (3) group different variables together*. In a model of COVID-19 hospitalizations, it may seem like a mere conventional choice whether to draw two graphs for people over and under 60, respectively, choose 70 as the cutoff age, or draw just one combined graph. After all, these are just different ways to represent the same data. Yet having a cutoff age makes it more likely that decision makers who rely on the model will make age-differentiated policies, which will use the model cutoff age as their anchor.

Another example concerning variable individuation and grouping: Israeli Jewish society is composed of two major ethnic groups: Ashkenazim ("westerners") and Mizrahim ("easterners"). It's commonly thought that these categories emerged in the 1950s, when Israel experienced mass immigration of Jews from Arab countries. But as Anat Leibler's (2014) archival research reveals, this binary ethnic classification emerged in a scientific context in the 1940's, prior to Israel's establishment. Demographer, statistician, and Zionist activist Roberto Bachi (1909–1995), who would later become the first director of the Israel Central Bureau of Statistics, was worried about the lower demographic growth rate of the Jewish population compared with that of the local Palestinian population. Within the statistical data at his disposal, Bachi correlated country of birth with a low/high average number of births per woman, respectively. By doing so, he claimed to have identified two distinct ethnic groups: Ashkenazim and Mizrahim. His explicit motivation was to identify a group that could make Jews win the demographic race with the Palestinians.

As Leibler (2014) notes, "social or ethnic categories neither have their roots in primordial objective reality nor emerge spontaneously" (272). It was not an inevitable fact that the Israeli Jewish population would be seen as consisting of exactly two major ethnicities grouped this way. But once this binary classification acquired foothold, it had far-reaching implications. Nowadays this classification is taken for granted. For Jewish Israelis, "Ashkenazim" and "Mizrahim" constitute central identity categories, which are strongly correlated with socioeconomic status and political affiliation. Their contingent scientific origin is largely unknown.

Value-laden decisions about what and how to represent may lead to dire consequences when the representations are used as *diagnostic tools*, specifically as discriminators between normal and abnormal cases. For example, from the 1950s and until at least the 1990s, studies of coronary heart disease included almost only white male subjects. Consequently, the known risk factors and symptoms associated with coronary heart disease were those typical of white men, which are different from those typical of women. Many women were consequently fatally misdiagnosed and mistreated (Riska 2010).

Commonly used pulse oximeters, which measure blood oxygen saturation, and have been widely used during the coronavirus pandemic, constitute another example of flawed diagnostic tools based on biased representation decisions. Such devices have been found to underestimate the severity of the condition of patients with darker skin, compared with lighter skin (Valbuena, Merchant, and Hough 2022). Similarly, commonly used spirometers are more likely to fail to detect emphysema in Black patients than in White ones (Liu

Representation Decisions

Social values affect representation decisions in scientific models.

Decisions about what to represent	Decisions about how to represent
- What entities to include in the model?	- On which data sources to rely? - How to individuate variables? - How to parse a variables space? - How to group different variables together?

Possible hazards
- incomplete knowledge
- irrelevant knowledge
- distracting knowledge
- error, when – in violation of scientific standards – an entity lacks properties required for extrapolating from the model to other target systems
- error, when an inadequate representation is used as a diagnostic tool

When a model is used outside its scientific context, its representation decisions may be socially entrenched to form rigid social categories.

Figure 3 Representation decisions

et al. 2022). As I argue elsewhere (Miller 2021a), these are also examples of how material technological artifacts, which some philosophers regard as a form of knowledge (Baird 2004; Humphreys 2009; compare Miller 2021a, 74–75), may embed social values.

"How to represent" decisions may overlap with "what to represent" decisions. Decisions about sampling, for example, are decisions about how to represent a large population using a smaller set of statistical points. But a sampling decision is also about what makes it into the model, hence it is about what to represent. Elisabeth Lloyd describes a similar case. A researcher who was working on the evolution of female orgasm in stumptail macaques set up his recording equipment such that the recording of the *female* orgasm was triggered by an increased heart rate of the *male* monkey. Lloyd (1993) writes:

> When I pointed out that the vast majority of female stumptail orgasms occurred during sex among the females alone, he replied that yes, he knew that, but he was only interested in the *important* orgasms. (142; emphasis in the origin)

This was both a decision about what to represent, namely, only female orgasms experienced with a male partner, and how to represent; namely, taking this set of recordings as a representative one. Describing this decision as value laden doesn't seem far-fetched (see Figure 3 for the section summary).

5.3 Social Values Fill the Gap of Underdetermination of Theory by Evidence

According to the thesis of underdetermination of theory by evidence, any body of evidence can be logically accommodated by more than one theory and perhaps infinitely many. It's argued that values "fill the gap" between theory and evidence; namely, because evidence alone doesn't determine theory or fix belief, agents must implicitly or explicitly appeal to values to choose which theory to accept or belief to form. They adopt the theories or beliefs most consonant with the values they cherish.[46]

Several distinctions are commonly drawn regarding underdetermination. One is between transient and in-principle underdetermination. Transient underdetermination regards a theory's being underdetermined by evidence as relative to a specific time. Transient underdetermination proponents argue that there is no guarantee that an underdetermined theory at a specific time will continue to be so as time goes by and more evidence is gathered (Laudan and Leplin, 1991). By contrast, proponents of in-principle underdetermination, associated with Quine (1951) and Duhem (1954), argue that in principle, evidence alone cannot fix a single theory regardless of the state of knowledge at a given time (Potter 1996; Longino 2016). Transient underdetermination assigns a more restricted role to values in theory choice, which may be diminished or eliminated at a future time (Norton 2008). Yet, there can be nontrivial conceptual differences between transiently underdetermined rival theories (Carrier 2011). Moreover, there is no guarantee that new relevant evidence *will* become available in the time to come.

Another distinction is between concrete and potential underdetermination. In concrete underdetermination, the evidence cannot fix a theory out of a set of concrete alternatives, while in potential underdetermination, it cannot fix a theory out of a set that consists inter alia of unconsidered theories. Potential underdetermination fuels sceptical arguments about scientific knowledge and objectivity, which state that if scientists choose a theory from a pool of all false or biased theories, and there are possible preferable yet unconsidered alternatives, then the truth or objectivity of scientific knowledge is questionable (Stanford 2010; Okruhlik 1994).

In their gap-filling role, values may influence theory choice bottom-up or top-down. Bottom-up, values implicitly affect scientists' background assumptions

---

[46] The prefix "under" has two meanings: "beneath" as in "underground," and "insufficient for," as in "underqualified." The meaning of "under" in "underdetermination" is the second; namely, "evidence underdetermines theory" means that the evidence is *insufficient* for fixing a choice of just one theory.

on which they base their theories. Such background assumptions are "neither self-evident nor logically true" (Longino, 2002, 128). Bottom-up influence accords well with potential underdetermination, because scientists' background assumptions restrict the possible theories they consider. Top-down, values act as "tiebreakers," namely, given two or more empirically equivalent theories, values determine which one is adopted. For example, *ceteris paribus*, if scientists adhere to simplicity, they choose the simpler theory. Bottom-up influence accords well with concrete underdetermination, because values help choose between actually available alternatives.

There is a debate about the role of values in filling the underdetermination gap, which concerns the *kinds* of values that legitimately fill the gap, and the *extent* values can, do, and should fill it. Regarding the kinds of values that legitimately fill the gap, some argue that only cognitive values, for example, simplicity, scope, or explanatory power, may legitimately influence theory choice. Such values are considered benign and internal to science (McMullin, 1983; Laudan, 2004). Others argue that social, political, and ideological values may legitimately fill the gap as well. For example, Kourany (2010, 69–75) argues that just like simplicity or scope, racial equality may constitute a legitimate reason for preferring a theory consonant with it rather than one that is not. This view is contested. Critics argue that the fact that social values play a role in theory choice doesn't entail that they should. According to this objection, social values reflect our desired social order – how the social world should be, while theories describe how the world is. Whether a theory is epistemically justified is a different question from whether it reflects a desired social order, therefore, the underdetermination gap-filling role of social values is illegitimate (Intemann 2005).[47]

Proponents of a legitimate gap-filling role for social values respond that the preceding objection presupposes an untenable, sharp, principled, and meaningful distinction between cognitive and social values. But social and cognitive values cannot be sharply distinguished, and since the critics acknowledge that the influence of *some* values, namely, cognitive values, on theory choice is necessary and benign, then social values may also play a legitimate role in theory choice (Longino 2002, 77–96; Machamer & Douglas 1999; Solomon 2001, 51–63). This doesn't mean, however, that the effect of any social value is legitimate in any context.

Another criticism of the gap argument concerns the *extent* social values fill the gap. John Norton (2008) writes that the gap argument rests on an "improvised and oversimplified account of the nature of inductive inference" (19). He

---

[47] Intemann (2005; 2015) acknowledges other legitimate roles for social values in science besides filling the underdetermination role.

argues that under most commonly used confirmation models, when two theories accommodate the same evidence, they usually don't enjoy the same inductive support, that is, the same evidence doesn't equally confirm them. Usually, the evidence inductively favors one theory over another. Norton therefore argues that social values fill the underdetermination gap only in rare cases in which theories enjoy similar inductive support.

The Strong Programme in the sociology of knowledge heavily relies on the underdetermination thesis. It holds that because theory is underdetermined by data, data alone don't explain why scientists adopt one theory rather than another. The Strong Programme also assumes that all scientists are exposed more or less to the same empirical reality: "reality is, after all, a common factor in all the vastly different cognitive responses that men produce to it. Being a common factor it's not a promising candidate to field as an explanation of that variation" (Barnes and Bloor 1982, 34). Explanation of theory choice or belief formation, according to the Strong Programme, should therefore focus on social values and interests. Within her framework of Social Empiricism, Miriam Solomon (2001) challenges this assumption. By reviewing several cases from the history of science, she argues that individual scientists are usually exposed to disparate pieces of the empirical puzzle. The empirical world looks differently from different perspectives, thus empirical reality isn't a common factor that can be neglected in explaining scientists' different theory choices. Given the Symmetry Thesis (Barnes and Bloor 1982, 20; Solomon 2001, 117), which states, roughly, that explanation should be attentive of the fact that people may form right beliefs for wrong reasons, and wrong beliefs for right reasons, it follows that a full explanation of scientific theory choice should address *all* factors, social and empirical, that affect an individual scientist's theory choice, as well as their aggregate effect on the community that is composed of many such individuals, or so Solomon argues (see Figure 4 for the section summary).

5.4 Social Values Adjust Evidential Weights

Values influence evidential reasoning by adjusting evidential weights in three ways:

(1) Values affect trust in testimony, specifically the degree of trust scientists give their peers.
(2) Values affect the evidential thresholds required for making justified epistemic judgments.
(3) Values affect the assignment of relative weights to different types of evidence within a body of discordant multimodal evidence.

> **Underdetermination of Theory by Evidence**
> Any body of evidence can be logically accommodated by more than one theory and perhaps infinitely many.
>
> ("underdetermine" = "be insufficient for determining")
>
> **Underdetermination:**
>
> | Transient | In principle |
> |---|---|
> | Relative to the state of evidence at a specific time | Independent of the state of evidence at a specific time |
> | **Concrete** | **Potential** |
> | Between existing alternative theories | Between unconceived theories |
>
> **The Gap-Filling Thesis:**
> Because evidence alone doesn't determine theory, agents must appeal to values to choose which theory to accept; they adopt the theories most consonant with the values they cherish.
>
> **Top-down**
> Values serve as tie-breakers between theories that accommodate the same data.
>
>
>
> **Bottom-up**
> Values affect theoretical background assumptions.
>
> **Debates**
> - What types of values fill or ought to fill the underdetermination gap: only cognitive values or also social values?
> - To what extent do social values affect theory choice by filling the underdetermination gap, and to what extent is theory choice constrained by empirical data and rules of inductive inferences?

Figure 4 Underdetermination of theory by evidence

This section reviews these three ways and argues that the role of values in adjusting evidential weights is distinct from their role in filling the gap of underdetermination of theory by evidence.[48]

Determining the degree of trust we give to another person's testimony is an epistemic evidential judgment. As Miranda Fricker (2007, ch. 1) argues, we draw inter alia on social stereotypes to determine how much trust to award to different speakers' testimonies. The stereotypes on which recipients of

---
[48] For an extended version of the argument in this section, see Miller (2014a).

testimony rely may be reliable or unreliable. When a stereotype is reliable, it correctly describes the majority of cases that fall under it. When stereotypes are unreliable, they negatively affect the person's testimonial beliefs by biasing them over time. When the same unreliable stereotypes are prevalent in society, common knowledge becomes systematically biased. In such a case, common knowledge represents the views of people to whom the prevailing stereotypes give credibility excess, and views of people who suffer from credibility deficit are dismissed.

Kristina Rolin (2004) argues that unreliable social stereotypes, which reflect prevailing sexist and racist values, aren't uncommon in science. Such social values skew scientists' assessment of their colleagues' trustworthiness. Sexist values make some male scientists unjustifiably underrate female scientists' testimonies and overrate males' testimonies, which biases both individuals' beliefs and the common body of scientific knowledge.

To some extent, then, values affect the degree of trust we give to others' testimony. Does the underdetermination model fully explain how they do so? If, as some epistemologists argue (Lipton 2007; Thagard 2005), the only way we decide whether to trust a person's testimony is by coming up with an explanatory theory about the testifier's trustworthiness, or by coming up with several such theories and choosing between them, then the underdetermination model fully explains the role of values in awarding trust.

But there are reasons to think that this model of trust in testimony is at least incomplete. Namely, even if theory choice is involved in some types of cases of awarding trust to testimony, there are cases in which it is not involved or is only partly involved, and they are still influenced by values. In such cases, the influence of value cannot be filling the gap of underdetermination of theory by evidence, because there is no theory involved, hence no gap to fill.

Three reasons lead me to argue that trust in testimony doesn't always involve, or only partly involves theory choice. *First*, even according to Thagard (2005) and Lipton (2007) who developed the theory-choice model of trust in testimony, a recipient of testimony switches to theory-choice mode only when she's triggered to suspect the testimony. It's possible that racist people are more easily triggered by a testimony of a person of color than nonracist people are, hence values affect them before the stage in which they come up with a theory about the speaker's trustworthiness. *Second*, we sometimes trust or mistrust because we *feel* that someone is trustworthy or nontrustworthy. Trust, at least sometimes, is an emotional stance (Lahno 2001; Fricker 2007, 75–80). The theory-choice model of trust doesn't address this emotional dimension. It's plausible that social values affect our emotional dispositions to trust. *Third*, according to the theory-choice model, we decide to trust after we form a belief that a person is trustworthy. But belief formation is involuntary, while trust, at least sometimes, seems voluntary. For example, in the game in which you let yourself fall back, there is a moment at which you voluntarily decide whether to trust your partner to catch you.

The second way in which values adjust evidential weights is by setting evidential thresholds. Sometimes we decide whether to trust evidence in the same way we decide whether to trust a person's testimony. For instance, Fricker (2007, 34–35) describes blind referees for a journal who are prejudiced against a new methodology rather than against a person. They resist the evidence because of some countervailing motivational investment, such as loyalty to the old methods, or fear of intellectual innovation.[49] For the same reasons as in the case of testimony, the influence of values on trust in evidence in such a case isn't necessarily an aspect of filling the underdetermination gap. I characterize this effect as adjusting evidential thresholds, because the stronger the evidence, the harder it is for a person to simply dismiss it.

Two additional examples illustrate how values adjust evidential thresholds and why their doing so isn't an aspect of filling the underdetermination gap. The first example is distinguishing signal from noise. The distinction between signal and noise isn't always sharp. Social values and personal interests and motivations may affect how scientists distinguish signal from noise. For example, a researcher who is under stress to publish or very committed to a certain theory may see signal where another scientist would see noise. Scientific theory helps researchers recognize a signal. It tells them what patterns in the data to expect. But theory doesn't necessarily guide the process of removing noise. Removing noise is done by checking and calibrating the instruments (Hacking 1983, 265) or by cleaning the data after an experiment or observation. Theory doesn't purport to explain the raw data in its entirety (Brown 1994, 129; Woodward 1989, 397; Frigg and Hartmann 2012), therefore removing certain data points from the raw data as noise doesn't require theoretical justification, hence the influence of values on its not filling the gap of underdetermination.

The second example of how values participate in setting evidential thresholds is the choice of threshold values of statistical significance. Statistical studies use mathematical metrics to evaluate evidence. One such metric, known as significance level (α) or critical p-value, is, roughly speaking, the

---

[49] For method bias, see Currie and Avin (2019).

acceptable threshold for the probability of wrongly accepting a false hypothesis. Wilholt (2009, 98) characterizes such metrics as conventional standards that impose implicit constraints on acceptable error probabilities within a research community. According to Wilholt (2009), "the standards adopted are arbitrary in the sense that there could have been a different solution to the same coordination problem, but once a specific solution is socially adopted, it's in a certain sense binding" (98). These values, however, are arbitrary only to a certain extent and within a certain range. The conventional critical p-value could have been 6 or 4.6 percent. Such values would also have served as reasonable solutions to the community's coordination problem. A critical p-value of 45 percent, however, would not have worked, as it would have meant that the community accepted as statistically significant results that were just slightly higher than chance.

We can identify two levels of influence of social values on setting threshold statistical values: the individual and the community. At both levels, their influence isn't restricted to filling the underdetermination gap. At the individual scientist's level, values, such as personal or ideological investments, may bias their judgment and cause them to infringe an explicit or implicit conventional standard by lowering or raising the evidential thresholds in a way that increases the likelihood of arriving at their preferred result, and violates their community's shared understanding of these thresholds (Wilholt 2009, 99).

At the community level, values may influence the conventional threshold values themselves. Different social values in different scientific contexts may participate in raising or lowering conventional threshold values. Such change of communal conventions has nothing to do with filling the underdetermination gap. For example, in significance testing, there is an inherent mathematical trade-off between minimizing false positives and false negatives. Values influence the balance between false positives and false negatives. The existing scientific standards, which are manifested inter alia in the widespread choice of the 5 percent significance level, are conservative in that they regard false positives as more serious errors than false negatives (Wilholt 2009, 99). As a conventional standard, the choice of a critical p-value isn't related to any specific scientific theory. It's used across the board in many sciences. Therefore, the effect of social values on p-value isn't a manifestation of underdetermination gap filling.

In some social contexts, scientists may adopt lower evidential thresholds, while in others they may adopt higher thresholds. In a society where smoking is customary and pervasive, scientists may set higher thresholds for evidence about the dangers of smoking than in a society that disapproves of smoking.

Ceteris paribus, the same evidence for the dangers of smoking may be considered insufficient in a smoking-friendly society and sufficient in a smoking-disapproving society. This may happen without a need arising in the smoking-friendly society to come up with rival theories to explain away the evidence, as the underdetermination model requires. When social values and norms change, evidential thresholds may change accordingly. If it becomes less socially acceptable to smoke for whatever reasons, the evidential thresholds may drop accordingly, again without a need arising for any theoretical justification for this drop. This example makes it clear that *Douglas' indirect role for values (Section 2.4) is one of the ways in which values adjust evidential thresholds.*

The third way in which values adjust evidential weight is that values affect relative weighing of discordant evidence. In science, multimodal evidence, namely, evidence from multiple techniques for the same theory, is often discordant. There are two types of discordance: inconsistency – an apparent contradiction between the hypotheses the evidence supports, and incongruity – different results that were produced under different background assumptions and are reported in "different languages" using different, possibly incommensurable units (Stegenga 2009).

There is no algorithmic or universally agreed method to combine multimodal evidence. Qualitative methods, for example, literature review, require judgment. Even quantitative methods, like meta-analysis, require judgment in their correct application at two stages: choosing the relevant evidence to begin with, and choosing the amalgamation method. Different inputs to the same method may produce different outputs. Different amalgamation methods, for example, different meta-analysis methods, may also produce different outcomes for the same evidence (Douglas 2012; Miller 2013, 1310–1311; Stegenga 2011).

In the face of multimodal discordant evidence, we may thus ask: Which evidence is more relevant to the given case? What relative weight should be given to each type of evidence? Which evidence should be more trusted?

Just as values lower or raise, to some extent, evidential thresholds, they also decrease or increase, to some extent, the relative weight an individual or a group assigns to different types of evidence within a body of multimodal discordant evidence. Given the same body of evidence, different persons or groups that adhere to different values may assign different relative weights to different evidence, without any need to theoretically justify the different weighing, namely, without values filling the underdetermination gap.

The section identified three roles social values play in evidential justification, which are distinct from filling the gap of underdetermination of theory by

> **Evidential Weight Adjusting**
> - Social values influence evidential reasoning by adjusting evidential weights in three ways:
>   (1) Values (social stereotypes) affect trust in testimony, specifically the degree of trust scientists give their peers.
>   (2) Values affect the evidential thresholds required for making justified epistemic judgments.
>   - This way corresponds to Douglas' indirect role for values (section 2.4).
>   - Examples:
>   - Values influence the distinction between signal and noise;
>   - Values affect thresholds of statistical significance;
>   (3) Values affect the assignment of relative weights to different types of evidence within a body of discordant multimodal evidence.
> - Values adjust evidential weights at the level of the individual researcher and at the level of the scientific community.
> - The influence of social values on epistemic evidential judgments isn't unconstrained, but within a range of possible weights.
> - Theory choice isn't necessarily involved in setting evidential weights, hence the role of values in adjusting evidential weights is different from their role in filling the gap of underdetermination of theory by evidence.

Figure 5 Evidential weight adjusting

evidence. First, values affect trust in testimony. Second, values lower and raise evidential thresholds. Third, values affect the relative weighing of multimodal discordant evidence. The significance of this argument is twofold. First, criticism of the underdetermination model which strives to show that the role of social values in evidential reasoning is modest doesn't apply to their role in adjusting evidential weights. Specifically, Douglas' argument from inductive risk, which refers to values' setting evidential threshold, is immune to the objections to the underdetermination model discussed in the previous section. Second, Douglas' argument from inductive risk is just *one* manifestation out of the possible three in which values adjust evidential weights. Philosophers of science and STS scholars may expand their analysis of case studies to attend to the two other ways (see Figure 5 for the section summary).

5.5 Social and Epistemic Elements Legitimate Each Other

In principle, any claim to knowledge can be indefinitely contested: there is no limit to how much inquiry, evidence, arguments, and deliberation is needed to justify a knowledge claim. According to Lorraine Code (1993), this means that there needs to be a *social* authority that declares a matter as settled, effectively nullifying any further objections. Traditionally, those already in power have had this authority, typically white men, and they have presented the views they certify as knowledge as if they were made from a neutral, objective perspective, though no human can occupy such a perspective (see section 3.6).

According to Jasanoff (2011), in liberal democracies, the public acceptance of content as actionable knowledge depends on the perceived political legitimacy of the political process of its adoption. What processes count as legitimate greatly varies between political cultures. Miller and Pinto (2022) argue, from epistemic and political normative perspective, that the social process of knowledge generation should be carried out such that all relevant qualified members of the epistemic community, including stakeholders and uncredited experts, can equally influence the knowledge generation process and its outcomes.

That knowledge is a social status has been recognized in feminist epistemology, history, and sociology of scientific knowledge (e.g., Code 1993; Collins and Pinch 1998; Galison 1987; Kusch 2002; Longino 2002), but has been overlooked or resisted in analytic epistemology. As Code (1993) argues, due to its roots in empiricism and positivism, analytic epistemology has focused on analyzing self-contained thought experiments describing a socially dislocated individual subject who forms a perceptual belief about a visible object in his vicinity. In such cases, a need doesn't arise to piece together different, potentially incompatible, partial subjective perspectives and declare a matter as settled, thus the social dynamics that are involved in it can be ignored.

Scientists are often autonomous and have the exclusive authority to reach closure and deem a content as known. When most members of the scientific community are persuaded that the weight of evidence supports a hypothesis or a model, they treat it as known (see Section 3.4). In such cases, it may wrongly look as if the weight of evidence alone transforms such content into knowledge. But *first*, as Henderson (2011) argues, to reach the status of knowledge, a theory or a model must pass gatekeepers, such as peer reviewers, journal editors, and grant committees. When a gatekeeper makes evidential judgments, she also exercises social authority when she allows or disallows content to pass the gate, for example to allocate funds to a research project, or to accept a paper for publication. While funding bodies, governmental and private, usually don't certify finished research, by making value judgments about which projects to fund, they indirectly affect institutional decisions about faculty hiring, equipment purchasing, degree programs, and overall research trajectory, thereby they influence the landscape of scientific knowledge from its inception (Sanders and Robison 1992).[50] *Second*, epistemic gatekeepers don't just make evidential judgments, they also make value judgments about inductive risks (Section 3.4). *Third*, they also check that the research has employed procedures such as double-blinding and declaration of conflicts of interest. Such social procedures have epistemic rationales, but someone with authority still needs to certify that they have been performed for the relevant research to gain knowledge status. *Fourth*, sometimes not all the members of the scientific community agree that the weight of evidence sufficiently supports a theory. In such cases, to reach closure, scientists implicitly or explicitly decide that the matter is settled and there is no more need to engage in evidential debates.

It becomes more visible that content must be socially or politically legitimated to be elevated to the status of scientific knowledge when scientists' authority to do so isn't exclusive. For example, as mentioned in Section 3.3.3, California law requires products that contain certain substances to bear a warning "this product can expose you to chemicals which are known to the State of California to cause cancer." The political authority of the state of California deems the content as known. Nutrition scientists resented the criteria for inclusion in the list of harmful substances and regarded the state as infringing proper expert scientific standards (Jasanoff 1996, 401–403). Namely, California challenged and revoked scientists' political authority – which is often exclusive, unnoticed, and taken for granted – to autonomously certify scientific knowledge.

Another example is the US Supreme Court *Daubert* (1993) decision and subsequent decisions that followed it. In *Daubert*, the Court devised criteria for the admissibility of scientific expert testimony that were said to examine whether the evidence was generated by methods deemed acceptable to scientists. Crucially, the court made judges, rather than scientists (or juries), the arbitrators of when these scientificity criteria are met (Jasanoff 1996, 403–406). In practice, the legal doctrine that developed following *Daubert* draws on, but differs from scientists' own standards. Some types of scientific research, which are mainly used as evidence in courts, must revise their methods to be admissible in court (Miller 2016).

In *Daubert*, there has been an osmosis of epistemic standards between science and the law because putative scientific knowledge claims have been subject to two social authorities that certify claims as scientific knowledge by

---

[50] For historical studies of science funding, see Oreskes (2021) and Solovey (2021). For alternative funding schemes see Avin (2019), and Shaw (2023). For private funding, see Holman and Elliott (2018) and Fernández Pinto (2022).

appealing to different, though related, epistemic standards. In such cases, epistemic elements and political elements colegitimate each other. The fact that expert evidence meets scientific standards strengthens the legitimacy of legal decisions based on it, and the fact that certain research is legally admissible as evidence, rather than inadmissible, strengthens its scientific standing.

We increasingly delegate or partly delegate to algorithms the authority to legitimate content as knowledge. Search engines, including academic search engines such as Google Scholar, and algorithmically generated scroll feeds in social media give more prominence to certain views, while other views are given less prominence or hidden away (Simon 2010a; Miller and Record 2013; Miller and Record 2017). There are different ways algorithms may generate epistemic closure, of which some give more room to human discretion and some less (Simon 2010b). There is ample room for research for social epistemology to study which algorithmic closure mechanisms are appropriate in different scientific and nonscientific contexts.

It might be objected that for every context, there is an objectively correct weighing of inductive risks, which is somehow metaphysically determined by the relevant facts of the case at hand. For example, given the different risks to producers and consumers, there exists an objective level of knowledge-level justification that a knowledge claim about a certain substance being carcinogenic needs to meet, regardless of what the State of California says. Note that this objection doesn't by itself deny Pragmatic Encroachment, but only that the weighing of inductive risks that sets the justification threshold for knowledge must be legitimated by a political authority.

It might further be objected that whether a content *constitutes* knowledge and whether it's *correctly identified* as knowledge by political actors are two separate questions. Whether it constitutes knowledge depends on whether it meets a certain standard of justification, regardless of whether it's politically recognized as knowledge. According to this objection, when an epistemic gatekeeper allows some content to pass an epistemic gate, they are merely identifying and reporting that it meets certain epistemic standards, but not endowing the content with a social status; namely, their speech act is declarative rather than constitutive.

I think these objections are wrong, but I am sympathetic to them. Indeed, in Section 3.6, I have myself argued that there are objective facts, possibly unbeknownst to the subjects, which determine whether a content they endorse passes the *Objective Justification* condition for scientific knowledge. I don't think, however, that this is the case regarding weighing of inductive risks or regarding undergoing some required social-epistemic procedures. As I have argued, it's conceptually clear and often discoverable in retrospect what the objective facts are that determine whether a content passes the *Objective Justification* condition: they are unconsidered relevant alternatives that explain the available evidence. It remains mysterious what these facts are in the case of weighing inductive risks and undergoing social-epistemic practices.

Still, I can see why this objection would appeal to those who are inclined to epistemic purism (the view that only truth-related factors determine what constitutes knowledge) or moral objectivism (the view that there are objective, universal moral standards). If you find this objection plausible, you may think about the political or social legitimation of knowledge as an extra step that putative knowledge must undergo in regulatory and policy contexts. Crucially, such political legitimation is still required for content that constitutes – in some metaphysical sense – scientific knowledge to *function in practice* as scientific knowledge (see Section 3.4).

It might still be objected knowledge and politics should be disjoint because political legitimation necessarily contaminates scientific knowledge. For example, from 1929 to 1956, the USSR officially adopted Trofim Lysenko's (1898–1976) false theory of biological inheritance, developed agricultural practices based on it, and persecuted biologists who dared to oppose it. The outcome was a setback to Soviet biological research, and vast famine. According to this objection, the lesson from this example is that the state should keep its hands off science. Indeed, maybe state noninterference is the right model in many cases, but this doesn't mean that politics and science can or should always be separated. The intervention of external political actors can also be epistemically beneficial. For example, cross-examination of forensic expert witnesses in court has revealed major flaws in their techniques (Jasanoff 1995, 55–57).

Additionally, as I have argued in this section, even when science is left to its own devices, scientific knowledge generation is a political process. Academic and scientific politics is still politics (some would even argue that it's more malicious than state politics). Scientists aren't nonpolitical creatures who merely impartially implement objective epistemic procedures. Political failures, such as inadequate representation or silencing of women or people of color, lead to epistemic flaws (recall the examples in Section 4). To repair these flaws, the political dimension of scientific knowledge generation should be acknowledged rather than suppressed.

As the Lysenko example illustrates, the fact that a content is politically certified as scientific knowledge from a descriptive sociological perspective doesn't necessarily mean that it constitutes scientific knowledge from an

> **The Political Legitimation of Knowledge**
> - Since the evidence required for reaching knowledge-level justification is potentially endless, social or political authority is required to declare a matter as settled and known.
> - When scientists are autonomous to decide what constitutes scientific knowledge, it's easy to miss that a content gains the status of scientific knowledge by virtue of scientists' social and political authority to award it with this status, rather than by the weight of evidence alone.
> - When other political actors, such as the court or the state, are also involved in the certification of scientific knowledge, the political legitimation of knowledge is easier to notice.
> - Different political actors or political cultures may have different epistemic standards and procedures that are required for political legitimation.
> - When scientists aren't the only political actors, they need to negotiate their epistemic standards and procedures with those of the other actors.
> - In our technological society, we increasingly delegate the role of political legitimation of knowledge to algorithms.

Figure 6 The political legitimation of knowledge

epistemic normative perspective, which still requires that it meet normative conditions for scientific knowledge, such as the ones I proposed in Section 3, which include both evidential requirements, and normative political requirements, such as a democratic decision-making processes and adequate representation of relevant views and stakeholders. In their analyses of empirical case studies, philosophers of science, social epistemologists, and science studies scholars can help us understand what meeting such normative standards for knowledge would mean in practice (see Figure 6 for the section summary).

References

Adam, Matthias. 2007. "Two Notions of Scientific Justification." *Synthese* 158(1): 93–108.
Anderson, Elizabeth. 2011. "Democracy, Public Policy, and Lay Assessments of Scientific Testimony." *Episteme* 8(2): 144–164.
Annis, David B. 1978. "A Contextualist Theory of Epistemic Justification." *American Philosophical Quarterly* 15(3): 213–219.
Avin, Shahar. 2019. "Centralized Funding and Epistemic Exploration." *The British Journal for the Philosophy of Science* 70(3): 629–656.
Baird, Davis. 2004. *Thing Knowledge: A Philosophy of Scientific Instruments*. Berkeley, CA: University of California Press.
Ballantyne, Nathan. 2018. "Epistemic Trespassing." *Mind* 128(510): 367–395.
Ballantyne, Nathan, Jared B. Celniker, and David Dunning. 2022. "Do Your Own Research." *Social Epistemology*, 1–16. doi: https://doi.org/10.1080/02691728.2022.2146469.
Barnes, Barry, and David Bloor. 1982. "Relativism, Rationalism and the Sociology of Knowledge." In Martin Hollis and Steven Lukes, eds., *Rationality and Relativism*, 1–20. Oxford: Blackwell.
Bateman, Grant A., and Brett D. Napier. 2011. "External Hydrocephalus in Infants: Six Cases with MR Venogram and Flow Quantification Correlation." *Child's Nervous System* 27(12): 2087–2096.
Beatty, John. 2006. "Masking Disagreement among Experts." *Episteme* 3(1–2): 52–67.
Beldecos, Athena, Sarah Bailey, Scott Gilbert, Karen Hicks, Lori Kenschaft, Nancy Niemczyk, Rebecca Rosenberg, Stephanie Schaertel, and Andrew Wedel. 1988. "The Importance of Feminist Critique for Contemporary Cell Biology." *Hypatia* 3(1): 61–76.
Betz, Gregor. 2013. "In Defence of the Value-Free Ideal." *European Journal for the Philosophy of Science* 3(2): 207–220.
Biddle, Justin B. 2014. "Can Patents Prohibit Research? On the Social Epistemology of Patenting and Licensing in Science." *Studies in History and Philosophy of Science* 45(1): 14–23.
Biddle, Justin, and Anna Leuschner. 2015. "Climate Skepticism and the Manufacture of Doubt: Can Dissent in Science Be Epistemically Detrimental?" *European Journal for Philosophy of Science* 5(3): 261–278.
Bird, Alexander. 2022. *Knowing Science*. Oxford: Oxford University Press.

Blachowicz, James. 2009. "How Science Textbooks Treat Scientific Method: A Philosopher's Perspective." *The British Journal for the Philosophy of Science* 60(2): 303–344.

Bogen, James. 1985. "Traditional Epistemology and Naturalistic Replies to Its Skeptical Critics." *Synthese* 64(2): 195–224.

Borgerson, Kirstin. 2011. "Amending and Defending Critical Contextual Empiricism." *European Journal for Philosophy of Science* 1(3): 435–449.

Brocas, Isabelle, and Juan D. Carrillo. 2008. "The Brain as a Hierarchical Organization." *American Economic Review* 98(4): 1312–1346.

Brown, James R. 1994. *Smoke and Mirrors: How Science Reflects Reality*. London: Routledge.

Brown, Mark B. 2009. *Science in Democracy: Expertise, Institutions, and Representation*. Cambridge, MA: Massachusetts Institute of Technology Press.

Brown, Matthew J. 2020. *Science and Moral Imagination: A New Ideal for Values in Science*. Pittsburgh, PA: University of Pittsburgh Press.

Buzzell, Andrew, and Regina Rini 2023. "Doing Your Own Research and Other Impossible Acts of Epistemic Superheroism." *Philosophical Psychology* 36(5): 906–930.

Carrier, Martin. 2011. "Underdetermination as an Epistemological Test Tube: Expounding Hidden Values of the Scientific Community." *Synthese* 180(2): 189–204.

Chakravartty, Anjan. 2017. "Scientific Realism." *The Stanford Encyclopedia of Philosophy*, ed. Edward N. Zalta. plato.stanford.edu/archives/sum2017/entries/scientific-realism.

Chakravartty, Anjan. 2022. "Scientific Knowledge vs. Knowledge of Science: Public Understanding and Science in Society." *Science & Education* 32: 1795–1812. doi: https://doi.org/10.1007/s11191-022-00376-6.

Chisholm, Roderick. 1977. *Theory of Knowledge*, 2nd ed. Englewood Cliffs, NJ: Prentice Hall.

Code, Lorraine. 1993. "Taking Subjectivity into Account." In Linda Alcoff and Elizabeth Potter, eds., *Feminist Epistemologies*, 15–48. New York: Routledge.

Collins, Harry M. 2010. "Elective Modernism." sites.cardiff.ac.uk/harrycollins/files/2016/02/elective-modernism-4.doc.

Collins, Harry M. 2014. *Are We All Scientific Experts Now?* Cambridge, UK: Polity.

Collins, Harry M., and Trevor Pinch. 1998. *The Golem: What You Should Know about Science*. Cambridge, UK: Cambridge University Press.

Conee, Earl, and Richard Feldman. 2004. *Evidentialism: Essays in Epistemology*. Oxford: Oxford University Press.

Crane, Diana. 1969. "Fashion in Science: Does It Exist?" *Social Problems* 16(4): 433–441.

Currie, Adrian M., and Shahar Avin. 2019. "Method Pluralism, Method Mismatch and Method Bias." *Philosophers' Imprints* 19(13): 1–22.

Dang, Haixin. 2019. "Do Collaborators in Science Need to Agree?" *Philosophy of Science* 86(5): 1029–1040.

Daubert v. Merrell Dow Pharmaceuticals, Inc. 1993. 509 U.S. 579.

de Melo-Martín, Inmaculada, and Kristen Intemann. 2018. *The Fight against Doubt: How to Bridge the Gap between Scientists and the Public*. Oxford: Oxford University Press.

de Ridder, Jeroen. 2014. "Epistemic Dependence and Collective Scientific Knowledge." *Synthese* 191(1): 37–53.

de Ridder, Jeroen. 2019. "How Many Scientists Does It Take to Have Knowledge?" In Kevin McCain and Kostas Kampourakis, eds., *What Is Scientific Knowledge? An Introduction to Contemporary Epistemology of Science*, 3–17. New York, NY: Routledge.

Dellsén, Finnur. 2021. "Consensus versus Unanimity: Which Carries More Weight?" *The British Journal for the Philosophy of Science*. In press. www.journals.uchicago.edu/doi/10.1086/718273.

Dellsén, Finnur. Forthcoming. "Disagreement and Consensus in Science." In Maria Baghramian, J. Adam Carter, and Richard Rowland, eds., *The Routledge Handbook of Disagreement*. philsci-archive.pitt.edu/19724/1/Dissent%20Draft14%20AcceptedWeb.pdf.

Dickhaut, John. 2009. "The Brain as the Original Accounting Institution." *The Accounting Review* 84(6): 1703–1712. https://digitalcommons.chapman.edu/esi_pubs/33.

Douglas, Heather. 2000. "Inductive Risk and Values in Science." *Philosophy of Science* 67(4): 559–579.

Douglas, Heather. 2009. *Science, Policy, and the Value-Free Ideal*. Pittsburgh, PA: University of Pittsburgh Press.

Douglas, Heather. 2012. "Weighing Complex Evidence in a Democratic Society." *The Kennedy Institute of Ethics Journal* 22(2): 139–162.

Dragos, Chris. 2016. "Which Groups Have Scientific Knowledge? Wray vs. Rolin." *Social Epistemology* 30(5–6): 611–623.

Dragos, Chris. 2021. "Epistemic Autonomy and Group Knowledge." *Synthese* 198(7): 6259–6279.

Duhem, Pierre. 1954. *The Aim and Structure of Physical Theory*. Princeton, NJ: Princeton University Press.

Elliot, Kevin. 2022. *Values in Science*. Cambridge, UK: Cambridge University Press.

Fantl, Jeremy, and Matthew McGrath. 2011. "Pragmatic Encroachment." In Sven Bernecker and Duncan Pritchard, eds., *The Routledge Companion to Epistemology*, 558–578. New York, NY: Routledge.

Faulkner, Paul R. 2018. "Collective Testimony and Collective Knowledge." *Ergo* 5(4): 103–126.

Fernández Pinto, Manuela. 2014. "Philosophy of Science for Globalized Privatization: Uncovering Some Limitations of Critical Contextual Empiricism." *Studies in History and Philosophy of Science* 47:10–17.

Fernández Pinto, Manuela, and Daniel J. Hicks. 2019. "Legitimizing Values in Regulatory Science." *Environmental Health Perspectives* 127(3): 035001-1–035001-8.

Fernández, Pinto, Manuela. 2022. "Science and Industry Funding." In Inkheri Koskinen et al., eds., *Global Epistemologies and Philosophies of Science*, 164–173. Abingdon: Routledge.

Fisch, Menachem. 2017. *Creativity Undecided: Toward a History and Philosophy of Scientific Agency*. Chicago, IL: University of Chicago Press.

Foley, Richard. 2005. "Justified Belief as Responsible Belief." In Matthias Steup and Ernest Sosa, eds., *Contemporary Debates in Epistemology*, 313–326. Malden, MA: Blackwell.

Foley, Richard. 2012. *When Is True Belief Knowledge?* Princeton, NJ: Princeton University Press.

Fricker, Miranda. 2007. *Epistemic Injustice: Power and the Ethics of Knowing*. Oxford: Oxford University Press.

Frigg, Roman, and Hartmann, Stephan. 2020. "Models in Science." In *The Stanford Encyclopedia of Philosophy*, ed. Edward N. Zalta. plato.stanford.edu/archives/spr2020/entries/models-science/.

Fronek, Patricia, and Lynne Briggs. 2018. "Faking Participant Identity: Vested Interests and Purposeful Interference." *Research Ethics* 14(2): 1–5.

Galam, Serge. 2012. *Sociophysics: A Physicist's Modeling of Psycho-Political Phenomena*. New York, NY: Springer.

Galison, Peter. 1987. *How Experiments End*. Chicago, IL: University of Chicago Press.

Gardiner, Georgi. 2025. "Purpose in Nature: A Function-First Naturalistic Epistemology." In Josh DiPaolo and Luis Oliveira, eds., *Kornblith and His Critics*. Chichester, West Sussex: Wiley.

Gerken, Mikkel. 2019. "Pragmatic Encroachment on Scientific Knowledge?" In Matthew McGrath and Brian Kim, eds., *Pragmatic Encroachment in Epistemology*, 116–140. New York, NY: Routledge.

Gerken, Mikkel. 2022. *Scientific Testimony: Its Roles in Science and Society*. Oxford: Oxford University Press.

Gettier, Edmund L. 1963. "Is Justified True Belief Knowledge?" *Analysis* 23(6): 121–123.

Giere, Ronald N. 1988. *Explaining Science: A Cognitive Approach*. Chicago, IL: University of Chicago Press.

Giere, Ronald N. 2002. "Scientific Cognition as Distributed Cognition," In Peter Carruthers, Stephen Stitch, and Michael Siegal, eds., *The Cognitive Basis of Science*, 285–299. Cambridge, UK: Cambridge University Press.

Giere, Ronald N. 2003. "The Role of Computation in Scientific Cognition." *Journal of Experimental & Theoretical Artificial Intelligence* 15(2): 195–202.

Giere, Ronald N. 2012. "Scientific Cognition: Human Centered But Not Human Bound." *Philosophical Explorations* 15(2): 199–206.

Glosserman, Scott and Nic Hill. 2010. *Truth in Numbers? Everything, According to Wikipedia* (Documentary Film). San Diego, CA: Underdog Pictures.

Goldberg, Sanford C. 2010. *Relying on Others: An Essay in Epistemology*. Oxford: Oxford University Press.

Goldberg, Sanford C. 2018. *To the Best of Our Knowledge: Social Expectations and Epistemic Normativity*. Oxford: Oxford University Press.

Goldberg, Sanford. C. 2021. "What Epistemologists of Testimony Should Learn from Philosophers of Science." *Synthese* 199(5): 12541–12559.

Goldman, Alvin I. 1976. "Discrimination and Perceptual Knowledge." *Journal of Philosophy* 73(20): 771–791.

Goldman, Alvin I. 1988. "Strong and Weak Justification." *Philosophical Perspectives* 2: 51–69.

Goldman, Alvin I. 2002. "Knowledge and Social Norms." *Science* 296(June 21): 2148–2149.

Gopnik, Alison, and Andrew N. Meltzoff. 1998. *Words, Thoughts, and Theories*. Cambridge, MA: Massachusetts Institute of Technology Press.

Green, Adam. 2016. *The Social Contexts of Intellectual Virtue: Knowledge as a Team Achievement*. London: Routledge.

Gundersen, Torbjørn. 2018. "Scientists as Experts: A Distinct Role?" *Studies in History and Philosophy of Science* 69: 52–59.

Habgood-Coote, Joshua, and Fenner Stanley Tanswell. 2023. "Group Knowledge and Mathematical Collaboration: A Philosophical Examination of the Classification of Finite Simple Groups." *Episteme* 20(2): 281–307.

Hacking, Ian. 1983. *Representing and Intervening: Introductory Topics in the Philosophy of Natural Science*. Cambridge, UK: Cambridge University Press.

Hacking, Ian. 1999. *The Social Construction of What*. Cambridge, MA: Harvard University Press.

Hacking, Ian. 2007. Kinds of People: Moving Targets. *Proceedings of the British Academy* 151: 285–318.

Haraway, Donna. 1988. "Situated Knowledges: The Science Question in Feminism and the Privilege of Partial Perspective." *Feminist Studies* 14(3): 575–599

Hardwig, John. 1985. "Epistemic Dependence." *Journal of Philosophy* 82(7): 335–349.

Hardwig, John. 1991. "The Role of Trust in Knowledge." *The Journal of Philosophy* 88(12): 693–708.

Harvard, Stephanie, Eric Winsberg, John Symons, and Amin Adibi. 2021. "Value Judgments in a COVID-19 Vaccination Model: A Case Study in the Need for Public Involvement in Health-Oriented Modelling." *Social Science & Medicine* 286(114323): 1–6.

Harvard, Stephanie, and Eric Winsberg. 2022. "The Epistemic Risk in Representation." *Kennedy Institute of Ethics Journal* 32(1): 1–31.

Heersmink, Richard. 2016. "The Cognitive Integration of Scientific Instruments: Information, Situated Cognition, and Scientific Practice." *Phenomenology and the Cognitive Sciences* 15(4): 517–537.

Henderson, David. 2011. "Gate-Keeping Contextualism." *Episteme* 8(1): 83–98.

Hicks, Daniel J. 2021. "When Virtues are Vices: 'Anti-Science' Epistemic Values in Environmental Politics." *Philosophy, Theory, and Practice in Biology* 14: 1–30.

Hills, Thomas T., and Yoed N. Kenett. (2022). "Is the Mind a Network? Maps, Vehicles, and Skyhooks in Cognitive Network Science." *Topics in Cognitive Science* 14(1): 189–208.

Holman, Bennet, and Kevin C. Elliott. (2018). "The Promise and Perils of Industry-Funded Science." *Philosophy Compass* 13: e12544.

House of Commons Health and Social Care, and Science and Technology Committees. 2021. *Coronavirus: Lessons Learned to Date. Sixth Report of the Health and Social Care Committee and Third Report of the Science and Technology Committee of Session 2021–22. HC 92.* committees.parliament.uk/publications/7497/documents/78688/default.

Humphreys, Paul. 2009. "Network Epistemology." *Episteme* 6(2): 221–229.

Hutchins, Edwin. 1995. *Cognition in the Wild.* Cambridge, MA: Massachusetts Institute of Technology Press.

Intemann, Kristen. 2005. "Feminism, Underdetermination, and Values in Science." *Philosophy of Science* 72(5): 1001–1012.

Intemann, Kristen. 2009. "Why Diversity Matters: Understanding and Applying the Diversity Component of the National Science Foundation's Broader Impacts Criterion." *Social Epistemology* 23(3–4): 249–66.

Intemann, Kristen. 2015. "Distinguishing between Legitimate and Illegitimate Values in Climate Modeling." *European Journal for Philosophy of Science* 5: 217–232.

Jablonka, Eva, and Marion J. Lamb. 2014. *Evolution in Four Dimensions: Genetic, Epigenetic, Behavioral, and Symbolic Variation in The History of Life*. Cambridge, MA: Massachusetts Institute of Technology Press.

Jasanoff, Sheila. 1995. *Science at the Bar: Law, Science, and Technology in America*. Cambridge, MA: Harvard University Press.

Jasanoff, Sheila. 1996. "Beyond Epistemology: Relativism and Engagement in The Politics of Science." *Social Studies of Science* 26(32): 393–418.

Jasanoff, Sheila. 2004a. "The Idiom of Co-production." In Sheila Jasanoff, ed., *States of Knowledge: The Co-production of Science and Social Order*, 1–12. London: Routledge.

Jasanoff, Sheila. 2004b. "Ordering Knowledge, Ordering Society." In Sheila Jasanoff, ed., *States of Knowledge: The Co-production of Science and Social Order*, 13–45. London: Routledge.

Jasanoff, Sheila. 2004c. "What Inquiring Minds *Should* Want to Know." *Studies in History and Philosophy of Science* 35(1): 149–157.

Jasanoff, Sheila. 2011. "Cosmopolitan Knowledge: Climate Science and Global Civic Epistemology." In J. Dryzek, R. B. Norgaard, and D. Schlosberg, eds., *Oxford Handbook of Climate Change and Society*, 129–143. Oxford: Oxford University Press.

Jasanoff, Sheila. 2012. "The Practices of Objectivity in Regulatory Science." In Charles Camic, Neil Gross, and Michèle Lamont, eds., *Social Knowledge in the Making*, 307–338. Chicago, IL: University of Chicago Press.

Keller, Evelyn Fox. 2015. "Cognitive Functions of Metaphor in the Natural Sciences." *Philosophical Inquiries* 3(1): 113–132.

Keren, Arnon. 2013. "Kitcher on Well-Ordered Science: Should Science be Measured against the Outcomes of Ideal Democratic Deliberation?" *Theoria* 28(2): 233–244.

Keren, Arnon. 2015. "Science and Informed, Counterfactual, Democratic Consent." *Philosophy of Science* 82(5): 1284–1295.

Kitcher, Philip. 1992. "The Naturalists Return." *Philosophical Review* 101(1): 53–114.

Kitcher, Philip. 1994. "Contrasting Conceptions of Social Epistemology." In Frederick F. Schmitt, ed., *Socializing Epistemology: The Social Dimensions of Knowledge*, 111–135. Lanham, MD: Rowman and Littlefield.

Kitcher, Philip. 2001. *Science, Truth, and Democracy*. New York, NY: Oxford University Press.

Kitcher, Philip. 2011. *Science in a Democratic Society*. Amherst, NY: Prometheus Books.

Knorr-Cetina, Karin. 1999. *Epistemic Cultures: How the Sciences Make Knowledge*. Cambridge, MA: Harvard University Press.

Kornblith, Hilary. 1983. "Justified Belief and Epistemically Responsible Action." *Philosophical Review* 92(1): 33–48.

Kourany, Janet A. 2010. *Philosophy of Science after Feminism.* New York, NY: Oxford University Press.

Kuhn, Thomas S. 1970. *The Structure of Scientific Revolutions*, 2nd ed. Chicago, IL: University of Chicago Press.

Kuhn, Thomas S. 2000. *The Road Since Structure: Philosophical Essays, 1970–1993, with an Autobiographical Interview.* Chicago, IL: University of Chicago Press.

Kukla, Rebecca. 2010. "The Ethics and Cultural Politics of Reproductive Risk Warnings: A Case Study of California's Proposition 65." *Health, Risk & Society* 12(4): 323–334.

Kukla, Rebecca. 2012. "'Author TBD': Radical Collaboration in Contemporary Biomedical Research." *Philosophy of Science* 79(5): 845–858.

Kusch, Martin. 2002. *Knowledge by Agreement.* Oxford: Oxford University Press.

Lackey, Jennifer. 2016. "What is Justified Group Belief?" *Philosophical Review* 125(3): 341–396.

Lahno, Bernd. 2001. "On the Emotional Character of Trust." *Ethical Theory and Moral Practice* 4: 171–189.

Lakatos, Imre, and Alan Musgrave, eds. 1970. *Criticism and the Growth of Knowledge.* Cambridge, UK: Cambridge University Press.

Larson, Brendon. 2011. *Metaphors for Environmental Sustainability: Redefining our Relationship with Nature.* New Haven, CT: Yale University Press.

Laudan, Larry. 2004. "The Epistemic, the Cognitive, and the Social." In P. Machamer and G. Wolters, eds., *Science, Values, and Objectivity*, 14–23. Pittsburgh, PA: University of Pittsburgh Press.

Laudan, Larry, and Leplin, J. 1991. "Empirical Equivalence and Underdetermination." *The Journal of Philosophy* 88(9): 449–472.

Leibler, Anat. 2014. "Disciplining Ethnicity: Social Sorting Intersects with Political Demography in Israel's Pre-state Period." *Social Studies of Science* 44(2): 271–292.

Lemetyinen, Henna. 2012. "Language Acquisition." *Simply Psychology* (October 24). www.simplypsychology.org/language.html.

Levy, Arnon. 2020. "Metaphor and Scientific Explanation." In Arnon Levy and Peter Godfrey-Smith, eds., *The Scientific Imagination*, 280–303. Oxford: Oxford University Press.

Levy, Neil. 2022. "Do Your Own Research!" *Synthese* 200(5): 1–19.

Lewis, Dyani. 2022. "Why the WHO Took Two Years to Say COVID Is Airborne." *Nature* 604(7904): 26–31.

Lipton, Peter. 2007. "Alien Abduction: Inference to the Best Explanation and the Management of Testimony." *Episteme* 4(3): 238–251.

Liu, Gabrielle Y., Khan, Sadiya S., Colangelo, Laura A., Meza, Daniel, Washko, George R., Sporn, Peter H. S., Jacobs Jr, David R., Dransfield, Mark T., Carnethon, Mercedes R. and Kalhan, Ravi. 2022. "Comparing Racial Differences in Emphysema Prevalence among Adults with Normal Spirometry: A Secondary Data Analysis of the CARDIA Lung Study." *Annals of Internal Medicine*, in press. doi.org/10.7326/m22-0205.

Lloyd, Elisabeth A. 1993. "Pre-Theoretical Assumptions in Evolutionary Explanations of Female Sexuality." *Philosophical Studies* 69(2/3): 139–153.

Longino, Helen E. 2002. *The Fate of Knowledge*. Princeton, NJ: Princeton University Press.

Longino, Helen E. 2013. *Studying Human Behavior: How Scientists Investigate Aggression and Sexuality*. Chicago, IL: University of Chicago Press.

Longino, Helen E. 2016. "Underdetermination: A Dirty Little Secret?" *STS Occasional Papers no. 4*. London: Department of Science and Technology Studies, UCL.

Longino, Helen E. 2022. "What's Social about Social Epistemology?" *Journal of Philosophy* 119(4): 169–195.

Ludeman, Kate, and Eddie Erlandson. 2004. "Coaching the Alpha Male." *Harvard Business Review* 82(5): 58–67.

Lugg, Andrew. 1978. "Disagreement in Science." *Journal for General Philosophy of Science* 9(2): 276–292.

Machamer, Peter, and Heather Douglas. 1999. "Cognitive and Social Values." *Science & Education* 8: 45–54.

Magnus, P. D. 2018. "Science, Values, and the Priority of Evidence." *Logos and Episteme* 9(4): 413–431.

Martin, Emily. 1990. "Toward an Anthropology of Immunology: The Body as Nation State." *Medical Anthropology Quarterly* 4(4): 410–426.

Martin, Emily. 1991. "The Egg and the Sperm: How Science Has Constructed a Romance Based on Stereotypical Male-Female Roles." *Signs* 16(3): 485–501.

Martini, Carlo, and Mattia Andreoletti. 2021. "Genuine versus Bogus Scientific Controversies: The Case of Statins." *History and Philosophy of the Life Sciences* 43(110): 1–23.

McKenna, Robin. 2022. "Is Knowledge a Social Phenomenon?" *Inquiry* doi: https://doi.org/10.1080/0020174X.2022.2135823.

McMullin, Ernest. 1983. "Values in Science." In Peter D. Asquith and Thomas Nickles, eds., *PSA 1982 (Vol. 2)*, 3–28. East Lansing, MI: PSA.

Miller, Boaz. 2013. "When Is Consensus Knowledge Based? Distinguishing Shared Knowledge from Mere Agreement." *Synthese* 190(7): 1293–1316.

Miller, Boaz. 2014a. "Catching the WAVE: The Weight-Adjusting Account of Values and Evidence." *Studies in History and Philosophy of Science* 47: 69–80.

Miller, Boaz. 2014b. "Science, Values, and Pragmatic Encroachment on Knowledge." *European Journal for the Philosophy of Science* 4(2): 253–270.

Miller, Boaz. 2015. "Why (Some) Knowledge Is the Property of a Community and Possibly None of Its Members." *The Philosophical Quarterly* 65(260): 417–441.

Miller, Boaz. 2016. "Scientific Consensus and Expert Testimony in Courts: Lessons from the Bendectin Litigation." *Foundations of Science* 21(1): 15–33.

Miller, Boaz. 2019. "The Social Epistemology of Consensus and Dissent." In David Henderson, Peter Graham, Miranda Fricker, and Nikolaj Pedersen, eds., *The Routledge Companion to Social Epistemology*, 228–239. New York, NY: Routledge.

Miller, Boaz. 2021a. "Is Technology Value-Neutral?" *Science, Technology, & Human Values* 46(1): 53–80.

Miller, Boaz. 2021b. "When Is Scientific Dissent Epistemically Inappropriate?" *Philosophy of Science* 88(5): 918–928.

Miller, Boaz. Forthcoming. "On the Relations between Individual and Communal Scientific Knowledge."

Miller, Boaz, and Ori Freiman. 2020. "Trust and Distributed Epistemic Labor." In Judith Simon, ed., *The Routledge Handbook of Trust and Philosophy*, 341–353. New York, NY: Routledge.

Miller, Boaz, and Meital Pinto. 2022. "Epistemic Equality: Distributive Epistemic Justice in the Context of Justification." *Kennedy Institute of Ethics Journal* 32(2): 173–203.

Miller, Boaz, and Isaac Record. 2013. "Justified Belief in a Digital Age: On the Epistemic Implications of Secret Internet Technologies." *Episteme* 10(2): 101–118.

Miller, Boaz, and Isaac Record. 2017. "Responsible Epistemic Technologies: A Social-Epistemological Analysis of Autocompleted Web Search." *New Media & Society* 19(12): 1945–1963.

Nagel, Ernest. 1979. *Teleology Revisited and Other Essays in the Philosophy and History of Science*. New York, NY: Columbia University Press.

National Academies of Sciences, Engineering, and Medicine. 2019. *Reproducibility and Replicability in Science*. Washington, DC: The National Academies Press. doi: https://doi.org/10.17226/25303.

Nelson, Lynn H. 1993. "Epistemological Communities." In Linda Alcoff and Elizabeth Potter, eds., *Feminist Epistemologies*, 121–160. New York, NY: Routledge.

Nickles, Thomas. 2021. "Historicist Theories of Scientific Rationality." *The Stanford Encyclopedia of Philosophy*, ed. Edward N. Zalta. plato.stanford.edu/archives/spr2021/entries/rationality-historicist.

Noddings, Nel. 2015. *Philosophy of Education*, 4th ed. London: Routledge.

Norton, John. 2008. "Must Evidence Underdetermine Theory?" In Martin Carrier, Don A. Howard, and Janet Kourany, eds., *The Challenge of the Social and The Pressure of Practice: Science and Values Revisited*, 17–44. Pittsburgh, PA: University of Pittsburgh Press.

O'Connor, Cailin. 2023. *Modelling Scientific Communities*. Cambridge, UK: Cambridge University Press.

Okruhlik, Kathleen. 1994. "Gender and the Biological Sciences." *Canadian Journal of Philosophy* 24(sup1): 21–42.

Oreskes, Naomi. 2007. "The Scientific Consensus on Climate Change: How Do We Know We're Not Wrong?" In Joseph F. C. DiMento and Pamela Doughman, eds., *Climate Change: What It Means for Us, Our Children, and Our Grandchildren*, 65–99. Cambridge, MA: Massachusetts Institute of Technology Press.

Oreskes, Naomi. 2019. *Why Trust Science?* Princeton, NJ: Princeton University Press.

Oreskes, Naomi. 2021. *Science on a Mission: How Military Funding Shaped What We Do and Don't Know about the Ocean*. Chicago, IL: University of Chicago Press.

Oreskes, Naomi, and Erik M. Conway. 2010. *Merchants of Doubt: How a Handful of Scientists Obscured the Truth on Issues from Tobacco Smoke to Global Warming*. New York, NY: Bloomsbury.

Page, Scott E. 2017. *The Diversity Bonus: How Great Teams Pay Off in the Knowledge Economy*. Princeton, NJ: Princeton University Press.

Parker, Wendy S. 2020. "Model Evaluation: An Adequacy-for-Purpose View." *Philosophy of Science* 87(3): 457–477.

Pentland, Alex. 2014. *Social Physics: How Good Ideas Spread: The Lessons from a New Science*. New York, NY: Penguin.

Perini, Laura. 2005. "Visual Representations and Confirmation." *Philosophy of Science* 72(5): 913–926.

Perini, Laura. 2012. "Truth-Bearers or Truth-Makers?" *Spontaneous Generations* 6(1): 142–147.

Popper, Karl R. 1970. "Normal Science and Its Dangers." In Lakatos and Musgrave, 51–58.

Popper, Karl R. 1972. *Objective Knowledge: An Evolutionary Approach*. Oxford: Clarendon Press.

Potter, Elizabeth. 1996. "Underdetermination Undeterred." In Lynn H. Nelson, and Jack Nelson, eds., *Feminism, Science, and the Philosophy of Science*, 121–138. Dordrecht: Kluwer.

Prince-Gibson, Eetta. 2020. "Israel Appointed 23 Experts to Lead Its Coronavirus Exit Strategy: None Are Women." *Haaretz* (May 8). www.haaretz.com/israel-news/2020-05-08/ty-article/.premium/israel-appointed-23-experts-to-lead-its-coronavirus-exit-strategy-none-are-women/0000017f-e187-d568-ad7f-f3ef22110000.

Pritchard, Duncan. 2010. "Cognitive Ability and the Extended Cognition Thesis." *Synthese* 175(S1): 133–151.

Quine, Willard V. O. 1951. "Two Dogmas of Empiricism." *The Philosophical Review* 60: 20–43.

Radder, Hans. 2017. "Which Scientific Knowledge is a Common Good?" *Social Epistemology* 31(5): 431–450.

Ranalli, Brent. 2012. "Climate Science, Character, and the 'Hard-Won' Consensus." *Kennedy Institute of Ethics Journal* 22(2): 183–210.

Record, Isaac, and Boaz Miller. 2022. "People, Posts, and Platforms: Reducing the Spread of Online Toxicity by Contextualizing Content and Setting Norms." *Asian Journal of Philosophy* 1(2): 1–19. doi:https://doi.org/10.1007/s44204-022-00042-2.

Rehg, William, and Kent Staley. 2008. "The CDF Collaboration and Argumentation Theory: The Role of Process in Objective Knowledge." *Perspectives on Science* 16(1): 1–25.

Rehg, William, and Kent Staley. 2017. "'Agreement' in the IPCC Confidence Measure." *Studies in History and Philosophy of Modern Physics* 57: 126–134.

Riska, Elianne. 2010. "Coronary Heart Disease: Gendered Public Health Discourses." In Ellen Kuhlmann and Ellen Annandale, eds., *The Palgrave Handbook of Gender and Healthcare*, 158–171. London: Palgrave Macmillan.

Rolin, Kristina. 2004. "Why Gender Is a Relevant Factor in the Social Epistemology of Scientific Inquiry." *Philosophy of Science* 71(5): 880–891.

Rolin, Kristina. 2009. "Scientific Knowledge: A Stakeholder Theory." In Jeroen van Bouwel, ed., *The Social Sciences and Democracy*, 62–80. New York, NY: Palgrave Macmillan.

Roush, Sherrilyn. 2005. *Tracking Truth: Knowledge, Evidence, and Science*. Oxford: Oxford University Press.

Rudner, Richard. 1953. "The Scientist *qua* Scientist Makes Value Judgments." *Philosophy of Science* 20(1): 1–6.

Sanders, John T. and Wade L. Robison. 1992. "Research Funding and the Value-Dependence of Science." *Business and Professional Ethics Journal* 11(1): 33–50.

Scheman, Naomi. 1995. "Feminist Epistemology." *Metaphilosophy* 26(3): 177–190.

Shaw, Jamie. 2023. "Peer Review, Innovation, and Predicting the Future of Science: The Scope of Lotteries in Science Funding Policy." *Philosophy of Science* 90(5): 1297–1306. doi: https://doi.org/10.1017/psa.2023.35.

Shieber, Joseph. 2015. *Testimony: A Philosophical Introduction*. New York, NY: Routledge.

Silva, Paul. 2019. "Justified Group Belief Is Evidentially Responsible Group Belief." *Episteme* 16(3): 262–281.

Silverstein, Arthur M. 2016. "The Whig History of Science: An Exchange." *The New York Review of Books* (February 25). www.nybooks.com/articles/2016/02/25/the-whig-history-of-science-an-exchange.

Simon, Judith. 2010a. "The Entanglement of Trust and Knowledge on the Web." *Ethics and Information Technology* 12(4): 343–355.

Simon, Judith. 2010b. "A Socio-epistemological Framework for Scientific Publishing." *Social Epistemology* 24(3): 201–218.

Sismondo, Sergio. 1996. *Science Without Myth: On Constructions, Reality, and Social Knowledge*. New York, NY: SUNY Press.

Slater, Matthew H., Joanna K. Huxster, Julia E. Bresticker, and Victor LoPiccolo. 2020. "Denialism as Applied Skepticism: Philosophical and Empirical Considerations." *Erkenntnis* 85: 871–890.

Solomon, Miriam. 2001. *Social Empiricism*. Cambridge, MA: Massachusetts Institute of Technology Press.

Solomon, Miriam. 2015. *Making Medical Knowledge*. New York, NY: Oxford University Press.

Solomon, Miriam, and Alan Richardson. 2005. "A Critical Context for Longino's Critical Contextual Empiricism." *Studies in History and Philosophy of Science* 36: 211–222.

Solovey, Mark. 2021. *Social Science for What? Battles over Public Funding for the "Other Sciences" at the National Science Foundation*. Cambridge, MA: Massachusetts Institute of Technology Press.

Staley, Kent, and Aaron Cobb. 2011. "Internalist and Externalist Aspects of Justification in Scientific Inquiry." *Synthese* 182: 475–492.

Stanford, Kyle. 2010. *Exceeding our Grasp: Science, History, and the Problem of Unconceived Alternatives*. Oxford: Oxford University Press.

Stegenga, Jacob. 2009. "Robustness, Discordance, and Relevance." *Philosophy of Science* 76(5): 650–661.

Stegenga, Jacob. 2011. "Is Meta-Analysis the Platinum Standard of Evidence?" *Studies in History and Philosophy of Biological and Biomedical Science* 42(4): 497–507.

Stegenga, Jacob. 2016. "Three Criteria for Consensus Conferences." *Foundations of Science* 21(1): 35–49.

Stegenga, Jacob. 2018. *Medical Nihilism*. Oxford: Oxford University Press.

Stich, Stephen. 1990. *The Fragmentation of Reason: Preface to a Pragmatic Theory of Cognitive Evaluation*. Cambridge, MA: Massachusetts Institute of Technology Press.

Suppe, Frederick. 1993. "Credentialing Scientific Claims." *Perspectives on Science* 1(2): 153–203.

Swan, Melanie. 2015. "Blockchain Thinking: The Brain as a Decentralized Autonomous Corporation." *IEEE Technology and Society Magazine* 34(4): 41–52.

Thagard, Paul. 2005. "Testimony, Credibility, and Explanatory Coherence." *Erkenntnis* 63: 295–316.

Tollefsen, Deborah P. 2015. *Groups as Agents*. Cambridge, UK: Polity.

Turner, Stephen. 2001. "What Is the Problem with Experts?" *Social Studies of Science* 31(1): 123–149.

Valbuena, Valeria S. M., Raina M. Merchant, and Catherine L. Hough. 2022. "Racial and Ethnic Bias in Pulse Oximetry and Clinical Outcomes." *JAMA Internal Medicine* 182(7): 699–700.

Vickers, Peter. 2023. *Identifying Future-Proof Science*. Oxford: Oxford University Press.

Vines, John, Rachel Clarke, Peter Wright, John McCarthy, and Patrick Olivier. 2013. "Configuring Participation: On How We Involve People in Design." In *Proceedings of the SIGCHI Conference on Human Factors in Computing Systems* (CHI '13), 429–438. New York, NY: ACM.

Wagner, Roy. 2009. "Mathematical Marriages: Intercourse between Mathematics and Semiotic Choice." *Social Studies of Science* 39(2): 289–308.

Weinberger, David. 2011. *Too Big to Know: Rethinking Knowledge Now That the Facts Aren't the Facts, Experts Are Everywhere, and the Smartest Person in the Room Is the Room*. New York, NY: Basic Books.

Wilholt, Torsten. 2009. "Bias and Values in Scientific Research." *Studies in History and Philosophy of Science* 40: 92–101.

Woodward, James. 1989. "Data and Phenomena." *Synthese* 79: 393–472.

Wylie, Alison. 2003. "Why Standpoint Matters." In Robert Figueroa, and Sandra G. Harding, eds., *Science and Other Cultures: Issues in Philosophies of Science and Technology*, 26–48. London: Routledge.

Wynne, Brian. 1992. "Misunderstood Misunderstanding: Social Identities and Public Uptake of Science." *Public Understanding of Science* 1(3): 281–304.

Acknowledgments

This Element was supported by the Israel Science Foundation Grant 650/18 for the project "Skepticism about Testimony" and grant 984/23 for the project "Distrust in Science." I thank the Centre for Advanced Internet Studies (CAIS) for a visiting fellowship. Special thanks to Meital Pinto, my partner in life and in the academic race, for her support and endurance.

Cambridge Elements

Philosophy of Science

Jacob Stegenga
University of Cambridge

Jacob Stegenga is a Reader in the Department of History and Philosophy of Science at the University of Cambridge. He has published widely on fundamental topics in reasoning and rationality and philosophical problems in medicine and biology. Prior to joining Cambridge he taught in the United States and Canada, and he received his PhD from the University of California San Diego.

About the Series

This series of Elements in Philosophy of Science provides an extensive overview of the themes, topics and debates which constitute the philosophy of science. Distinguished specialists provide an up-to-date summary of the results of current research on their topics, as well as offering their own take on those topics and drawing original conclusions.

Cambridge Elements

Philosophy of Science

Elements in the Series

Values in Science
Kevin C. Elliott

Scientific Representation
James Nguyen and Roman Frigg

Philosophy of Open Science
Sabina Leonelli

Natural Kinds
Muhammad Ali Khalidi

Scientific Progress
Darrell P. Rowbottom

Modelling Scientific Communities
Cailin O'Connor

Logical Empiricism as Scientific Philosophy
Alan W. Richardson

Scientific Models and Decision Making
Eric Winsberg and Stephanie Harvard

Science and the Public
Angela Potochnik

Feminist Philosophy of Science
Anke Bueter

Abductive Reasoning in Science
Finnur Dellsén

The Social Dimensions of Scientific Knowledge: Consensus, Controversy, and Coproduction
Boaz Miller

A full series listing is available at: www.cambridge.org/EPSC

Printed in the United States
by Baker & Taylor Publisher Services